AF743942

PROPERTIES AND APPLICATIONS OF POLYMER DIELECTRICS

Edited by **Boxue Du**

Properties and Applications of Polymer Dielectrics
http://dx.doi.org/10.5772/63063
Edited by Boxue Du

Contributors

Mohd Hafizi Ahmad, Wei Shi, Junsheng Yu, Yifan Zheng, Abdul Naeem, Asad Mahmood, Tahira Mahmood, Youyuan Wang, Can Wang, Abdullahi Mas\'Ud, Ricardo Albarracin, Yunusa Umar, Manirul Haque, Jorge Ardila Rey, Boxue Du

Published by InTech
Janeza Trdine 9, 51000 Rijeka, Croatia

© The Editor(s) and the Author(s) 2017
The moral rights of the editor(s) and the author(s) have been asserted.
All rights to the book as a whole are reserved by InTech. The book as a whole (compilation) cannot be reproduced, distributed or used for commercial or non-commercial purposes without InTech's written permission. Enquiries concerning the use of the book should be directed to InTech's rights and permissions department (permissions@intechopen.com).
Violations are liable to prosecution under the governing Copyright Law.

(cc) BY

Individual chapters of this publication are distributed under the terms of the Creative Commons Attribution 3.0 Unported License which permits commercial use, distribution and reproduction of the individual chapters, provided the original author(s) and source publication are appropriately acknowledged. More details and guidelines concerning content reuse and adaptation can be found at http://www.intechopen.com/copyright-policy.html.

Notice

Statements and opinions expressed in the chapters are these of the individual contributors and not necessarily those of the editors or publisher. No responsibility is accepted for the accuracy of information contained in the published chapters. The publisher assumes no responsibility for any damage or injury to persons or property arising out of the use of any materials, instructions, methods or ideas contained in the book.

Publishing Process Manager Martina Usljebrka
Technical Editor SPi Global
Cover InTech Design team

First published May, 2017
Printed in Croatia
Legal deposit, Croatia: National and University Library in Zagreb

Additional hard copies can be obtained from orders@intechopen.com

Properties and Applications of Polymer Dielectrics, Edited by Boxue Du
p. cm.
Print ISBN 978-953-51-3147-2
Online ISBN 978-953-51-3148-9

We are IntechOpen, the world's leading publisher of Open Access books

Built by scientists, for scientists

3,700+
Open access books available

115,000+
International authors and editors

119M+
Downloads

151
Countries delivered to

Our authors are among the
Top 1%
most cited scientists

12.2%
Contributors from top 500 universities

WEB OF SCIENCE™

Selection of our books indexed in the Book Citation Index in Web of Science™ Core Collection (BKCI)

Interested in publishing with us?
Contact book.department@intechopen.com

Numbers displayed above are based on latest data collected.
For more information visit www.intechopen.com

For EU product safety concerns:
IN TECH d.o.o., Prolaz Marije Krucifikse Kozulić 3, 51000 Rijeka, Croatia,
info@intechopen.com or visit our website at intechopen.com.

Meet the editor

Boxue Du (M'00-SM'04) received his ME degree in Electrical Engineering from Ibaraki University and his PhD degree from Tokyo University of A&T. Since 2002, he has been a professor at the Department of Electrical Engineering, School of Electrical Engineering and Automation, Tianjin University, China. Professor Du has been active in polymer insulation research since the 1990s. His fundamental and meaningful work provided original and valuable information to the design, evaluation, and monitoring of electrical power devices for the purpose of enhancing the operating reliability in various environments. He is a member of IEEJ, senior member of CSEE, member at several WG in CIGRE, and associate editor of the *IEEE Transactions on Dielectrics and Electrical Insulation*.

Contents

Preface

In recent years, new insulating materials are in great need due to the complexity of the power system and the development of science and technology. There are many domestic and foreign scholars doing the related research now. Traditionally, inorganic substances such as mica and silicon dioxide were used as dielectric materials. However, with the rapid development of engineering dielectric, polymers are playing a more important role in dielectric materials. With the further research of polymer dielectrics, increasingly more principles need to be revised, and more problems that have not been well recognized are focused on.

As the polymer industry gradually turns into large-scale industry, polymers get more application, offering an alternative to the traditionally inorganic and ceramic material. This is due to their properties such as highly flexible, easier handling, good chemical stability, and customization for specific applications. Moreover, the engineering dielectrics, especially nanodielectrics, have been put into use in the fields of electrical energy, electronic communications, and military aviation. Their properties have scientific significance and engineering value. In this book, several important thoughts and information are put forward.

The book gives the reader an overview on electrical properties in Dr. Abdullahi's chapter. More knowledge on other applications such as converter transformer, transistor, and energy storage is provided in the section of polymer properties and application. Besides, this book also presents some recent researches on typical polymer material such as silicon rubber and LDPE. This may provide some clues of advanced polymer properties for both engineers and researches.

I have long been engaged in research work in the field of engineering dielectric. Polymer dielectrics started for me in/at/when and has become an important project of my lab in the last few years. It is a collection of our team's efforts. I wish to express a grateful acknowledgment to the authors of all these chapters. In addition, the cooperation of InTech is also highly recognized.

Boxue Du
Key Laboratory of Smart Grid of Ministry of Education
School of Electrical Engineering and Automation
Tianjin University
Tianjin, China

Section 1

Polymer Properties and Application

Chapter 1

Polymer Dielectric in Organic Field-Effect Transistor

Wei Shi, Yifan Zheng and Junsheng Yu

Additional information is available at the end of the chapter

http://dx.doi.org/10.5772/65916

Abstract

In this chapter, we aim to present an overview of the polymer dielectrics in organic field-effect transistors and their applications. In the first section, we give a short introduction of polymer dielectrics in organic field-effect transistors. We illustrate multilayer, hybrid, and cross-linked polymer dielectrics adopted in organic field-effect transistors. Then we introduce the available biomaterials engaged as polymer dielectrics in organic field-effect transistors. We mainly focus on the utilization of silk fibroin, DNA, and DNA base pair dielectrics. We end the chapter by presenting the applications of polymer dielectrics. We elaborate that the polymer dielectrics can function as the electrode buffer layer, as well as the organic field-effect transistor-based gas sensor, inverter, and memory.

Keywords: polymer dielectric, organic field-effect transistor, biomaterial, gas sensor

1. Introduction

Organic field-effect transistor (OFET) is an indispensable component in the field of organic electronics, which has been developed to realize low-cost, flexible large-area products, and biodegradable electronics [1–3]. Compared with the conventional silicon dioxide–based device, OFETs with polymer dielectrics are ideally compatible with flexible substrates and solution process [4]. Apparently, the solution processable polymer dielectric is very attractive, for it is compatible with spin-coating, casting, and printing at room temperature and under ambient conditions. Meanwhile, this capability has practical advantages when coupled with large-scale production using the patterning technique [5]. Moreover, as the function dependent on the structure of polymer dielectric is readily available, to design and synthesize a structure with certain function becomes feasible [6], which results in complementary kinds of polymer dielectrics. Recently, with more and more polymer biomaterials engaged in the OFETs to serve

INTECH
open science | open minds

© 2017 The Author(s). Licensee InTech. This chapter is distributed under the terms of the Creative Commons Attribution License (http://creativecommons.org/licenses/by/3.0), which permits unrestricted use, distribution, and reproduction in any medium, provided the original work is properly cited.

as the dielectric, the resource of polymer dielectric has become environmentally friendly and very broad. Due to such kinds of polymer dielectrics, OFETs are available in versatile applications including sensor, inverter, and memory [7].

2. Introduction of polymer dielectric in organic field-effect transistor

2.1. Polymer dielectric in organic field-effect transistor

In a common configuration, the OFET consists of the source and drain electrodes, semiconductor, dielectric, and gate. The functional part of an OFET device is the current channel, which exists in the first few monolayers of the semiconductor upon the dielectric. Therefore, the interface between the dielectric and the semiconductor plays a crucial role in the device performance, which makes the requirements for polymer dielectric material for OFET rather stringent [8, 9]. The crucial parameter of a dielectric material is the maximum possible electric displacement (D_{max}) the dielectric can sustain:

$$D_{max} = \varepsilon_0 k E_B \quad (1)$$

where ε_0 is the vacuum permittivity, k is the dielectric constant, E_B is the dielectric breakdown field, and the capacitance per area C_i, which is defined as:

$$C_i = \varepsilon_0 \frac{k}{d} \quad (2)$$

where d is the thickness of the dielectric. It is obvious that the capacitance magnitude is not only governed by the k value but also by the thickness of the dielectric.

The first detailed study of different polymer dielectrics in OFET was reported by Peng et al. in 1990 [10]. The OFET was fabricated on glass using five kinds of polymer dielectrics. They found that there was a strong correlation between the insulator's k value and the field-effect mobility. Then, in 1997, the first high performance plastic transistor was realized by Bao et al. [11]. They employed polyimide as the dielectric and all the essential components were printed directly on the plastic. The OFET had a field-effect mobility of 0.01–0.03 cm^2/Vs. In 2002, Klauk et al. made a step further achievement as the "all-polymer" circuit which integrated a 250 nm thick melamine cross-linked poly(vinyl pyrrolidone) (PVP) [12]. The cross-linked PVP with the capacitance of ~11–12 nF/cm^2 made the OFET yield a high carrier mobility of 3 cm^2/Vs. In the next year, Veres et al. reported that the interaction between the dielectric and the semiconductor plays a crucial role in the charge carrier transport. They found out that for a larger permittivity of the dielectric, the more charge carrier was localized at the surface of the dielectric [13]. Thus, there is a contradictory selection between the high dielectric material and the low permittivity material. In order to obtain such a balance, in 2004, Park et al. introduced double polymer layers as the dielectric in OFET [14]. This structure consisted of a thin PVP layer in contact with the semiconductor, which could induce good charge transport properties and a thick poly(vinyl acetate) (PVAc) layer as the bottom layer to realize good dielectric properties.

In recent years, the tendency of the research on polymer dielectrics has been toward printable, flexible, and biocompatible materials. In 2015, Huang et al. developed a versatile self-healing polymer blend dielectric without salts and integrated it into the OFET [15]. This high capacitance of polymer blend dielectric could even induce the healing of the functional layer coated above it. In 2016, Jung et al. introduced a kind of cross-linked poly(methyl methacrylate) (PMMA) in an n-type OFET, which examined the application of the OFET for flexible circuit [16]. Schmidt et al. fabricated a low-voltage fully printed flexible OFET using three layers of dielectric of CYTOP (low-k), PVA (intermediate), and P(VDF-TrFE-CTFE) (high-k) [17]. Moreover, more and more biomaterials, such as glucose, indigo, and nucleic acid–based materials, are used in the OFETs [18]. Among them, DNA and silk fibroin (SF) are two widely researched dielectric biomaterial in OFET. For example, Liang et al. had adopted DNA-hexadecyltrimethyl ammonium chloride (CTMA) as the dielectric layer in OFET to successfully realize a nonvolatile memory [19]. Wang et al. employed the SF as the gate dielectric in OFET and obtained a high mobility of 23.2 cm^2/Vs and a low operating voltage of -3 V [20].

2.2. Multilayer polymer dielectric

In OFETs, the bulk and interface properties of gate dielectric are both crucial in determining device performance. Therefore, two prerequirements have to be met for the dielectric material: (1) a high gate capacitance, which reduces the required gate voltage for sufficient charge accumulation in the channel, can be obtained by decreasing the film thickness or increasing the dielectric constant; (2) a trap-free interface between semiconductor and dielectric, which enhances the carrier mobility [17, 21]. A high k material has been considered as a good choice for the dielectric in OFET due to the excellent bulk insulating properties. However, the surface polarization of the high k material is much more higher than that of the low k material, which can significantly deteriorate the device performance. Therefore, to achieve high performance device, there should be a meticulous consideration between the high k and the low k materials. In this case, the utilization of the multilayer dielectric, combining a very thin low k layer at the interface to the semiconductor with a thick high k material for sufficient gate insulation is a reliable approach.

Yu et al. made an effective approach on multilayer dielectrics of PMMA and PVA in OFET [22]. Through analyzing the electrical characteristics of OFETs with various PVA/PMMA arrangements as shown in **Figure 1(a)**, it was found that one of the origins of the hysteresis was the trap in PVA bulk as well as at the interface of pentacene/PVA. Meanwhile, the results showed that the memory window was proportional to the amount of traps in PVA and the charge density at the interfaces of gate/PVA or PVA/pentacene. Then, the pentacene OFETs based on bilayer dielectrics of PMMA/SF was developed, as shown in **Figure 1(b)** [23]. The PMMA/SF bilayer dielectric exhibited a high field-effect mobility of 0.21 cm^2/Vs and a high current on/off ratio of 1.5×10^4. The performance enhancement was mainly attributed to the crystallization improvement of the pentacene and the smaller interface trap density at the SF/pentacene interface. Meanwhile, a low contact resistance also indicated that a good contact of electrode/organic was formed.

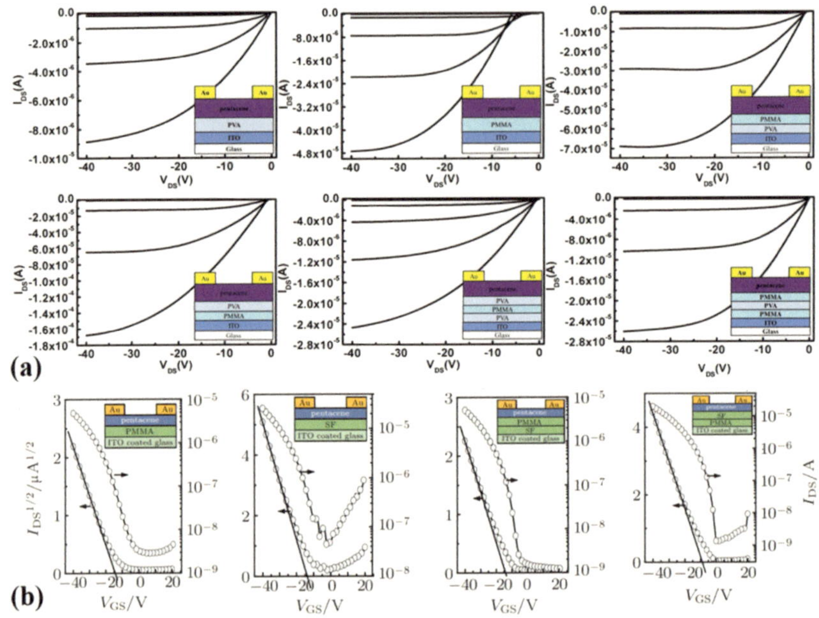

Figure 1. Configuration and performance of OFETs with (a) PVA and PMMA multi-layer dielectrics; (b) PMMA/SF bi-layer dielectrics [22, 23]. Copyright 2013, American Institute of Physics; Copyright 2014, IOP Publishing.

2.3. Hybrid polymer dielectric

An easily processable polymer typically has a low dielectric constant and good mechanical properties but requires large gate dielectric thicknesses due to high leakage current. One of possible solutions is to combine inorganic-organic or organic-organic materials as hybrid gate dielectrics. These complementary constituents ideally combine high permittivity of the inorganic inclusions with the high breakdown strength, mechanical flexibility, and easy processability of the organic counterparts [24]. Sun et al. explored a blending polymer of dielectric of polyphenyleneoxide (PPO) and polystyrene (PS) to enhance the performance of pentacene OFET [25], which could control both the inter-grain-enhancing process and the nucleation-controlling process. The optimized morphology of pentacene thus led to the enhancement of mobility to 3.6 cm^2/Vs. Yu and coworkers also employed the PMMA/zinc oxide (ZnO) hybrid as the dielectric for OFET [26]. The resulted morphology of the pentacene grown on the hybrid dielectric was responsible for the enhanced sensing performance to ammonia (NH_3) of the OFET.

Moreover, the polymer dielectric can not only serve as the dielectric layer in OFET but can also play a role in the semiconducting layer [27]. Yu et al. made it a step further to adopt the poly(3-hexylthiophene) (P3HT)/PS hybrid as the semiconductor [28]. The relationship between the molecular arrangement, aggregation, and charge transport in P3HT:PS blends with the boiling points and solubility in different solvents like chloroform (CF), o-xylene (XY), chlorobenzene (CB), and 1,2-dichlorobenzene (DCB) was systematically analyzed, as shown in **Table 1**. The result showed that DCB with the highest boiling temperature was beneficial to achieve distinct lateral aggregation of P3HT in the blend film. An optimized composition of 1.6 wt% P3HT in the PS matrix with three times increase of mobility and two times increase of current on/off ratio were obtained, compared to that of the pure P3HT (8 wt%). Li et al. adopted a blend of 6, 13-bis(triisopropylsilylethynyl)pentacene (TIPS-pentacene) and PS as the semiconducting layer [29]. They synthesized the PMMA and PS functionalized with both propargyl and azido groups by free radical copolymerization to serve as the dielectric to obtain a high performance OFET with a field-effect mobility of 0.59 cm^2/Vs, and an on/off current ratio of 10^5. Feng et al. fabricated a solution processed bottom-gate bottom-contact OFET, which could be able to sustain hybrid low-/high-voltage operation [30, 31]. In their devices, a channel engineering approach was used to obtain low-voltage operation, by inducing phase separation with a blend of TIPS-pentacene and PS to form an ultrathin high crystalline channel. Since the approach did not rely on enlarging the gate dielectric capacitance, the low-voltage OFET with a relatively thick dielectric layer was shown to be able to sustain high-voltage operation.

		μ (cm^2/V s)	V_T (V)	*SS* (V/dec)	On/off ratio (10^2)
DCB	P3HT	0.010 ± 0.002	-1.5 ± 0.2	10.0 ± 0.2	2.7 ± 0.1
	P3HT:PS	0.030 ± 0.002	0.2 ± 0.05	6.0 ± 0.2	6.2 ± 0.1
CB	P3HT	0.004 ± 0.0005	3.0 ± 0.2	9.5 ± 0.2	1.4 ± 0.1
	P3HT:PS	0.010 ± 0.002	4.0 ± 0.2	6.0 ± 0.2	2.9 ± 0.1
XY	P3HT	0.010 ± 0.002	0.5 ± 0.05	9.5 ± 0.2	2.1 ± 0.1
	P3HT:PS	0.005 ± 0.0005	7.5 ± 0.2	11.0 ± 0.2	2.7 ± 0.1
CF	P3HT	0.002±0.0005	-5 ± 0.2	15 ± 0.2	1 ± 0.1
	P3HT:PS	–	–	–	–

Table 1. Characteristics of pure P3HT and P3HT:PS blend OFETs based on different solvents [27]. Copyright 2015, Elsevier.

2.4. Cross-linked polymer dielectric

Commonly, there are two problems that exist in polymer gate dielectrics. One is the stability and the other one is the electrical robustness [29]. Meanwhile, when multilayer polymer dielectrics are applied, the elimination of the dissolution or swelling problem of the upper layer toward the under layer is desired. Cross-linking method can not only enhance solvent resistance and thermal stability, but also improve the electrical robustness of dielectric materials [32]. Moreover, the interference of each dielectric in the multilayer structure can be solved through the cross-link process. Commonly, the cross-linking in the dielectric layer can be achieved by photo or thermal reactions [33–35].

Li et al. synthesized the PMMA and PS functionalized with both propargyl and azido groups by free radical copolymerization, which could be effectively cross-linked by thermal azide-alkyne cycloaddition reaction at a relatively low temperature of 100°C as shown in **Figure 2** [29, 36]. This bifunctional approach significantly improved the efficiency of cross-linking reactions in the solid state and substantially enhanced the solvent resistance of the cross-linked dielectric layers. The OFET exhibited a high device performance with a field-effect mobility of 0.59 cm^2/Vs, and an on/off current ratio of 10^5 as mentioned above in the "hybrid polymer dielectric" part. At the same time, they synthesized two azide functionalized polymers by free radical copolymerization [37, 38]. Each new polymer was effectively cross-linked with a small molecule cross-linker by a thermally activated reaction at 100°C. This cross-linking method is compatible with plastic substrates for flexible electronic applications.

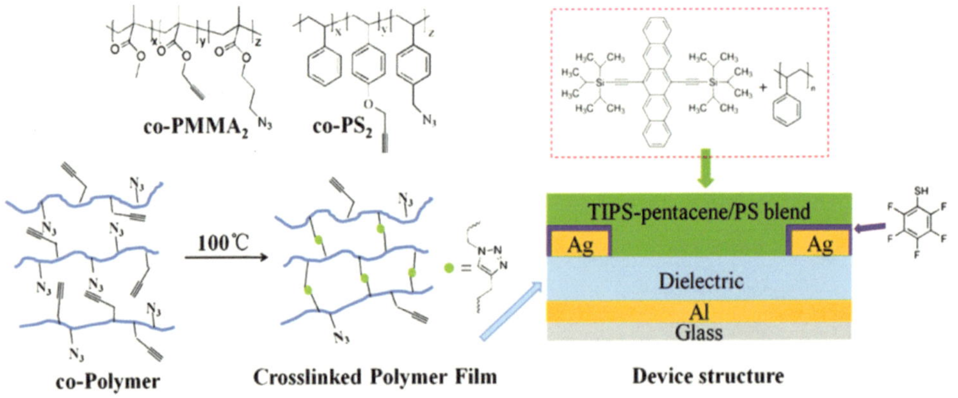

Figure 2. Schematic of the OFET and the chemical structures of the relevant materials [29, 36]. Copyright 2015, RSC Publishing; Copyright 2014, RSC Publishing.

From the above discussion, we can see that polymer dielectric is widely researched in OFET due to its inherent advantages of solution processability and flexibility. With the abundant resources, we can choose polymer dielectric with certain properties to serve different functions in hybrid and multilayer structures. Moreover, through cross-linking method, the atmosphere stability of polymer dielectric can be significantly enhanced, which brings a wider utilization for OFET and its electronic device.

3. Biomaterial-engaged polymer dielectric

3.1. Biomaterial in organic field-effect transistor

Nowadays, the research field of organic electronics is becoming more and more interdisciplinary, since biomaterial possesses superior attributes of chemical abundance, biodegradability, and low cost. OFETs are frequently used in the detection of biomaterials, whereas biomaterials are engaged in OFETs to serve as one of the functional layers as shown in **Figure 3** [18, 39–41].

Substrates made of caramelized glucose, edible hard gelatin, and commercially available plastics based on potato and corn starch provide examples for metabolizable or biodegradable substrates [18]. DNA and DNA base pairs have been used as gate dielectric in OFETs and memory elements [42–44]. Beta-carotene and indigo have been employed as the p-type and n-type materials in OFETs, exhibiting the charge mobility of ~10^{-4} cm^2/Vs [45, 46].

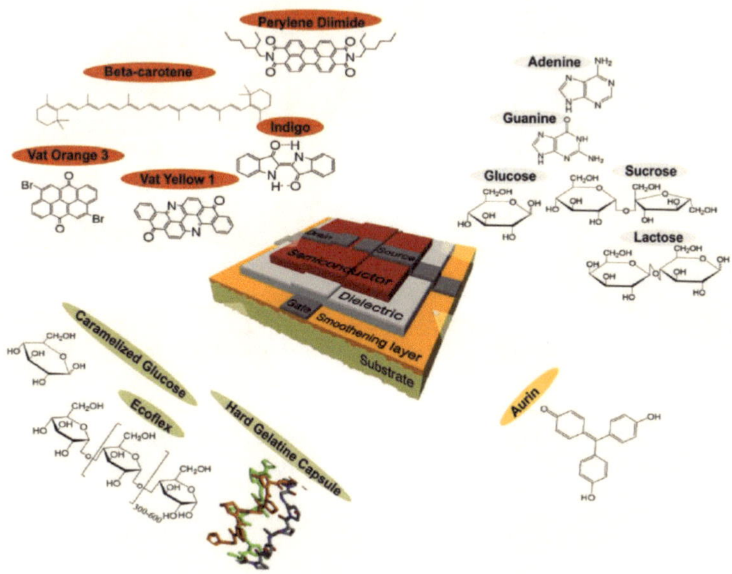

Figure 3. Utilization of natural materials or materials inspired by nature in OFETs [18]. Copyright 2010, Wiley.

3.2. Silk fibroin dielectric

Silk fibroin is one of the silk proteins emitted by the silkworm, which forms the structural center of silk with sericin around it. It is a natural biopolymer consisting of the repeated amino acid sequence of alternating glycine (gly) and alanine (ala) as shown in **Figure 4(a)** [47]. Until now, silk fibroin has been employed in various electronic fields, including the contact lenses [48], the platform for transistors [49], and photonic devices [50]. These applications are benefited from its unique characteristics of optical transparency, electrical insulation, and flexibility.

Yu et al. reported an enhanced performance pentacene OFETs consisting of PMMA/SF bilayer dielectric in 2014 [23]. The SF had good dielectric properties as shown in **Figure 4(b)**. The surface morphology of SF is very uniform with the root-mean-square surface roughness value of 1.3 nm as shown in **Figure 4(d)** (we can compare it with the smooth surface of PMMA shown in **Figure 4(c)**). The OFETs had a relatively high mobility of 0.21 cm^2/Vs with an enhanced on/off ratio of 1.5 × 10^4. This was mainly attributed to the crystallization improvement of the pentacene grown on SF (as shown in **Figure 4(e)** and **(f)**) and the smaller interface trap density at the SF/pentacene interface. Then, the utilization of SF-engaged OFET to function as the nitrogen dioxide (NO_2) sensor was further studied [51]. In this research, SF was deposited on the top of PMMA to act as the dielectric layer, resulting in an increase of NO_2 sensing performance.

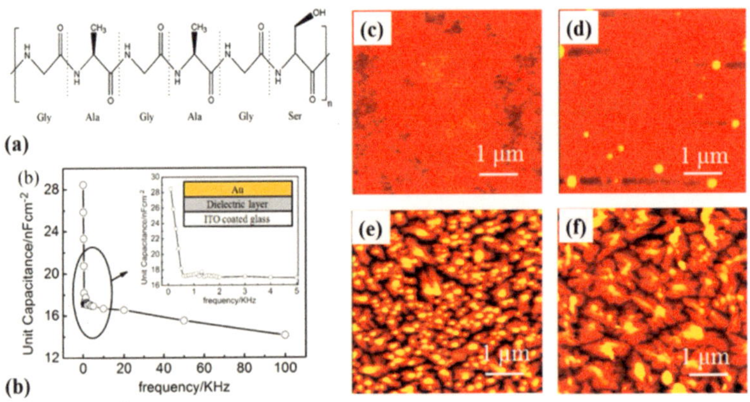

Figure 4. (a) Molecular structure of SF; (b) capacitance versus frequency (C-F) property of SF, the inset shows the C-F image in a frequency range of 0–5 KHz and structure of the device; AFM images of (c) PMMA; (d) SF; pentacene grown on (e) PMMA and (f) SF [23]. Copyright 2014, IOP Publishing.

3.3. DNA and DNA base pair dielectric

DNA is a complicated macromolecule consisting of four base pairs of guanine, adenine, thymine, and cytosine. Singh et al. first reported the utilization of DNA in OFET as a dielectric layer in 2006 [52]. They used a DNA-based biopolymer, derived from salmon milt and roe sac waste by-products for the gate dielectric, in which the current was modulated over three orders of magnitude using gate voltages less than -10 V. In most studies, since the purified DNA dissolves only in water, it should be modified through a cationic surfactant (hexadecyltrimethyl ammonium chloride, CTMA) cation exchange reaction to enhance solubility, processing, and stability.

Stadler et al. reported the utilization of DNA-CTMA as the gate dielectric in n-type methanofullerene as well as p-type pentacene-based OFETs working at low-voltage levels and low gate leakage currents [43]. They further realized a nonvolatile memory element based on the large hysteresis in the transfer characteristics of these DNA-based OFET. As DNA is soluble only in water, which is not compatible with most organic solvents, Yu et al. first introduced the spray coating method to fabricate DNA film. In 2016, spray-coated DNA on top of the PMMA dielectric was used to fabricate a DNA-functioned NO_2 sensor based on OFET as shown in **Figure 5(a)** and **(b)** [53]. The high-sensing performance is ascribed to the negatively-charged phosphate groups in DNA molecules, which can interact with NO_2 analytes as shown in **Figure 5(c)**.

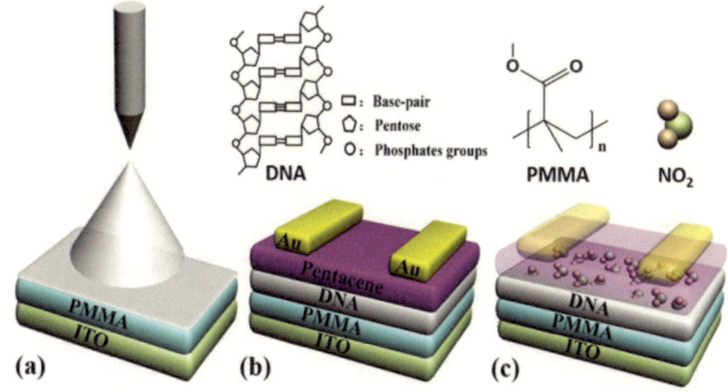

Figure 5. (a) Schematic representation of spray-coating technique; (b) device architecture of OFET sensor and the molecular structures of DNA and PMMA; (c) representation of the sensing mechanism of OFET-based NO_2 chemical sensor [53]. Copyright 2016, Elsevier.

Nature provides an overwhelming diversity of materials for human being. Hence, looking for natural or nature-inspired materials appears to be a promising route for the fabrication of fully biodegradable and biocompatible organic electronics. In the future work, to make the all biomaterial- based device, which is compatible with human body, is an attractive area of the biomaterial-engaged polymer dielectrics.

4. Application-related polymer dielectric

4.1. Electrode buffer layer in organic field-effect transistor

Dielectric polymers can not only serve as the dielectric layer in the OFETs, but can also be adopted as the modification layer at the interface between the semiconductor and the electrodes, due to their smooth and hydrophobic surface properties. Especially, the high performance bottom-contact OFET is required for its potential application in the large-scale industrialized production due to the remarkable advantage in the size-controlled fine lithography processing [54]. One of the major limits in fabricating high performance OFET in bottom-contact configuration is the large contact resistance at organic/electrode interface, which results in a pronounced current loss. This makes the employment of the electrode buffer layer necessary.

Yu et al. fabricated bottom-contacted OFET by using PMMA as an electrode buffer layer between P3HT layer and gold electrodes in OFETs and obtained a five-fold enhancement of whole mobility [55]. The relatively rough surface of gold (**Figure 6(a)**) is modified by PMMA (**Figure 6(b)**). The uniformity and hydrophobicity of PMMA surface were responsible for the remarkable reduction of contact resistance at P3HT/electrode interface while they enhanced the crystallinity of P3HT as shown in **Figure 6(c)**. Three polymer dielectrics of PMMA, PS, and polyvinylidene fluoride (PVDF) as the electrode buffer layers were further studied in top-contacted OFET to decrease the contact resistance (**Figure 6(d)**) [30]. All the OFETs incorporating with the buffer layers obtained a significant enhancement of the device performance, whereas the device employing PMMA exhibited the highest charge mobility of 0.59 cm^2/Vs. This was due to the optimal surface energy and appropriate dielectric constant of PMMA, which are favorable for the growth of pentacene crystal.

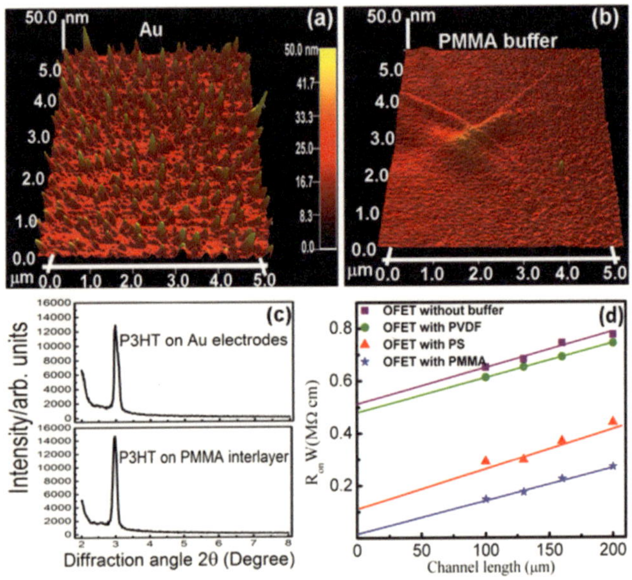

Figure 6. AFM images of (a) gold and (b) PMMA; (c) XRD analysis of P3HT; (d) contact resistance of OFETs [55]. Copyright 2013, American Institute of Physics.

4.2. Organic field-effect transistor-based gas sensor

OFET-based sensors have been intensively researched by virtue of their incomparable advantage of abundant material resource, physical flexibility, and elaborate array compatibility [56–59]. Moreover, OFET holds the unique capability of gate bias modulation, by which the signal can be easily amplified over orders of magnitude. This provides a promising future of OFETs for the ultra-low detection [60–64]. The controllable multiparameters of charge mobility, threshold voltage, and current on/off ratio make OFETs more suitable for selective detection [65]. Until now, OFET has been used to detect the solution, gas as well as the biomaterials [66].

Different kinds of polymer dielectrics perform differently in the OFET sensors. Yu et al. investigated the effect of various kinds of polymer dielectrics on the sensing performance of the OFET sensors. NH_3 gas sensors were fabricated based on pentacene OFETs using polymers including PVA, PVP, PMMA, or PS as the gate dielectric as shown in **Figure 7(a)** [67]. The OFETs with PS as the gate dielectric could achieve the detection limit as low as 1 ppm. Meanwhile, the recovery properties of OFETs with PS were also enhanced. The variation of the sensing performance of OFET sensors with different dielectrics was proved to be mainly induced by the different properties of dielectric/pentacene interfaces. Furthermore, the low-trap density of PS dielectric surface and the absence of polar groups in PS dielectric were responsible for the high performance of NH_3 sensors. Moreover, the OFET incorporating ZnO/PMMA hybrid dielectric was fabricated to detect NH_3 as shown in **Figure 7(b)** [26]. The OFETs exhibited a 23% current change under 75 ppm NH_3, as well as a remarkable shift of threshold voltage and field-effect mobility. The sensing mechanism was ascribed to the decreased grain size of pentacene formed on the ZnO/PMMA hybrid dielectric, facilitating NH_3 to diffuse into the conducting channel.

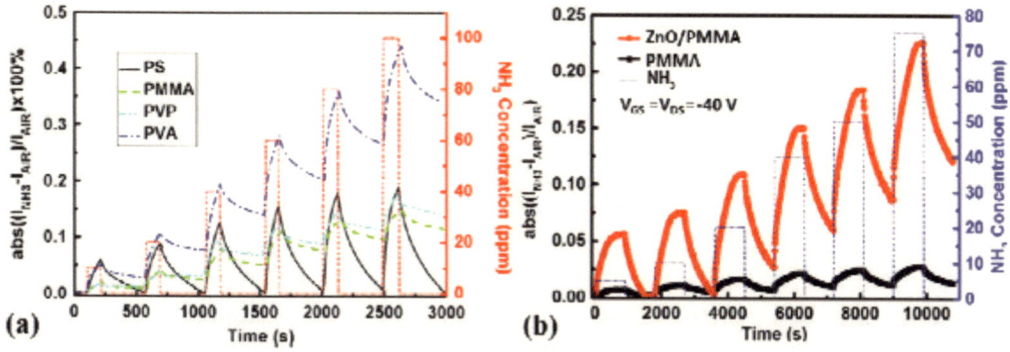

Figure 7. Real-time response curves of (a) OFET with PS, PMMA, PVP, and PVA as dielectric; (b) OFET with ZnO/PMMA and PMMA as dielectric exposed to NO_2 [26, 67]. Copyright 2014, Elsevier; Copyright 2013, Elsevier.

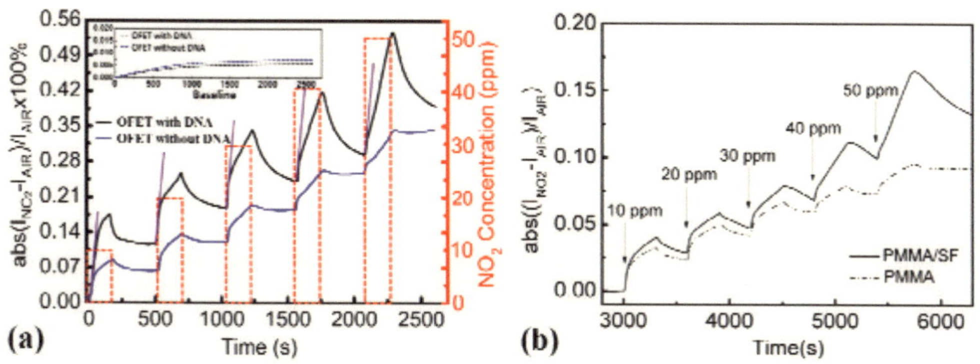

Figure 8. Real-time response curves of (a) OFET with and without DNA; (b) OFET with and without SF exposed to NO_2 gas in the concentrations ranging from 10 to 50 ppm [51, 53]. Copyright 2015, Springer; Copyright 2016, Elsevier.

In terms of the biomaterials discussed above, they also exhibit good sensing performance in OFETs. A DNA-based OFET sensor was realized for the detection of NO_2 as shown in **Figure 8(a)** [53]. In this sensor, DNA was introduced between the gate dielectric and the organic semiconductor via spray-coating to function as the detecting layer for NO_2 analyte. There were remarkable shifts of 14.7% in saturation current and 14.4% in charge carrier mobility after exposure to 10 ppm NO_2 analyte. With the concentration of NO_2 increased to 50 ppm, the shifts of 22.8% in saturation current and 16.6% in charge carrier mobility had been obtained. The sensing performance was ascribed to the negatively-charged phosphate groups in DNA molecules, which could interact with NO_2 analytes and lead to a superior sensing performance of OFET incorporating with DNA (**Figure 5(c)**). Silk fibroin was also employed on top of PMMA dielectric in OFETs to detect NO_2 as shown in **Figure 8(b)** [51]. When exposed to a high NO_2 concentration of 50 ppm, the saturation current of the OFETs exhibited an increase of 16%. The superior sensing performance was due to the interaction between the hydroxyl and the amidogen of the SF biomaterial and NO_2 molecules at the interface of dielectric/organic layer.

4.3. Organic field-effect transistor-based inverter

Inverters are the basic and indispensable parts in the case of electronic circuits. Especially, complementary inverters, which usually consist of both p-type and n-type transistors, are widely applied in silicon-based circuits. Besides, due to the performance inconsistency of p-type and n-type organic semiconductors, unipolar inverter, which can be functionalized with only one type of transistors, is extensively studied in the field of organic electronics [68, 69]. The OFETs with a significant controllable threshold voltage (V_{th}) are indispensable to construct the high performance unipolar inverter. Ultraviolet (UV)/ozone (UVO) treatment can modify the physical and chemical characteristics of the polymer surface, leading to a shift of the V_{th} without changing the properties of the bulk material [70, 71].

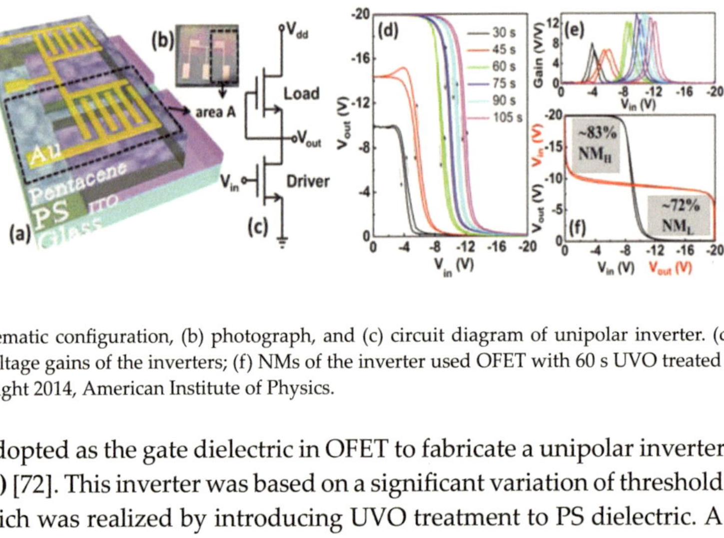

Figure 9. (a) Schematic configuration, (b) photograph, and (c) circuit diagram of unipolar inverter. (d) VTCs and (e) corresponding voltage gains of the inverters; (f) NMs of the inverter used OFET with 60 s UVO treated PS as load transistor [72]. Copyright 2014, American Institute of Physics.

PS was also adopted as the gate dielectric in OFET to fabricate a unipolar inverter as shown in **Figure 9(a)–(c)** [72]. This inverter was based on a significant variation of threshold voltage (V_{th}) of OFETs, which was realized by introducing UVO treatment to PS dielectric. A controllable V_{th} shift of more than 10 V was obtained in the OFETs by adjusting the treating time, and the unipolar inverters exhibited an inverting voltage near 1/2 driving voltage and a noise margin of more than 70% of ideal value as shown in **Figure 9(d)–(f)**. The dramatic controllable V_{th} of OFETs was attributed to the newly generated oxygen functional groups in the PS dielectric induced by UVO treatment. Guo et al. developed a novel organic/inorganic hybrid integration architecture to realize low-voltage complementary inverters with low temperature (not exceeding 150°C) solution-processed semiconductor and PVA dielectric layers [73]. The fabricated inverter had a voltage gain larger than 15 at an operation voltage of 3 V. The inverters were further fabricated on a polyethylene naphthalate plastic substrate with the bottom-gate bottom-contact configuration [28]. In the devices, UV cross-linked PVA was used as the gate dielectric layer and the devices show a high DC voltage gain up to 67.3 at a supply voltage of 3 V. Also, the inverter with the low-operation voltage based on a channel engineering approach was fabricated [74]. The relatively thick (even with a 400-nm thick) and low dielectric constant polymer dielectric of PVA could be used in the inverter.

4.4. Organic field-effect transistor-based memory

OFET-based memory mostly relies on the hysteresis of the dielectric. As to a nonvolatile transistor, a voltage is added between the gate electrode and the semiconducting channel. Through the charge storage, the effective gate voltage within the device then differs from the applied voltage. The resulted polarization phenomenon creates an additional electronic state in the device. PVA is a commonly adopted polymer dielectric in OFET-based memory. The hysteresis mechanism of OFETs with PVA dielectric can be concluded in two aspects: the charge transport in PVA bulk and the charge trapping/detrapping process in PVA bulk and/or at the interface of organic semiconductor/PVA.

The research on multilayer dielectric of PMMA and PVA in OFET was also carried out [22]. Through analyzing the electrical characteristics of OFETs with various PVA/PMMA arrangements, it was found that one of the origins of the hysteresis was the trap in PVA bulk as well as at the interface of pentacene/PVA. Meanwhile, the results showed that the memory window was proportional to the amount of traps in PVA and the charge density at the interfaces of gate/PVA or PVA/pentacene. Therefore, the memory window could be controlled to around 0–10 V by tuning the thickness and combination of triple-layer polymer dielectrics strategy. Meanwhile, as the dielectric interfaces also greatly influenced the number of densities and the mobility of charge carrier in adjacent semiconductors, the surface issue should also be taken into consideration. Feng et al. fabricated a novel type of OFET-based write-once read-many memory (WORM) device [75]. The device used an ultraviolet cross-linkable matrix dielectric polymer of poly (vinyl cinnamate) (PVC) mixed with ionic compounds 10-methyl-9-phenyl-acridinium perchlorate ($MPA^{+}ClO_4^{-}$) to form an ion-dispersed gate dielectric layer. Under an applied gate voltage bias, the migration of cations and anions in opposite directions formed space charge polarization in the gate dielectric layer, resulting in a change of electrical characteristics. Through UV illumination method to cross-link the matrix polymer, the stability of the formed space charge polarization could be enhanced. Hence, the OFET could function as a WORM with the applied voltage bias to define the polarization, and at the same time the UV illumination could stabilize the stored data.

There is a wide utilization of OFETs functionalized by polymer dielectric. The most common adopted application is the OFET-based sensors. As the effective current channel lies at the interface between the semiconductor and the polymer dielectric, to modify the surface of the dielectric becomes an effective way to enhance the sensing performance. Moreover, polymer dielectric can also play a role in the electrode buffer layer as well as the inverter and memory, which makes the polymer dielectrics not simple function as the dielectric, but also as a versatile material.

5. Summary

In this section, we make a concise introduction of polymer dielectric in OFETs, including the multilayer, hybrid, and cross-linked polymer dielectric. In addition, we had a detailed overview of the available biomaterials of the silk fibroin and the DNA and DNA base pair as polymer dielectrics. As one of the most important layer in OFETs, polymer dielectric possesses a promising application as interface modification layer in OFETs, as well as the functional layer in OFET-based sensor, inverter, and memory. Hence, the polymer dielectric holds versatile inherent properties to be explored and is paced toward the future in the field of organic electronics.

Acknowledgements

We acknowledge the National Natural Science Foundation of China (NSFC) (Grant no. 61177032 and No. 61675041), the Fundamental Research Funds for the Central Universities (Grant no. ZYGX2010Z004), the Foundation of Innovation Groups of NSFC (Grant no. 61421002), the China Scholarship Council (File no. 201506070085 and no. 201506070069), and Science & Technology Department of Sichuan Province (Grant no. 2016HH0027).

Author details

Wei Shi, Yifan Zheng and Junsheng Yu*

*Address all correspondence to: jsyu@uestc.edu.cn

State Key Laboratory of Electronic Thin Films and Integrated Devices, School of Optoelectronic Information, University of Electronic Science and Technology of China (UESTC), Chengdu, People's Republic of China

References

[1] Bittle, E. G.; Basham, J. I.; Jackson, T. N.; Jurchescu, O. D.; Gundlach, D. J., Mobility overestimation due to gated contacts in organic field-effect transistors. *Nature Communications* 2016, *7*, 10908.

[2] Zhang, C.; Chen, P.; Hu, W., Organic field-effect transistor-based gas sensors. *Chemical Society Reviews* 2015, *44* (8), 2087–2107.

[3] Sirringhaus, H., 25th Anniversary Article: Organic field-effect transistors: the path beyond amorphous silicon. *Advanced Materials* 2014, *26* (9), 1319–1335.

[4] Mannsfeld, S. C.; Tee, B. C.; Stoltenberg, R. M.; Chen, C. V. H.; Barman, S.; Muir, B. V.; Sokolov, A. N.; Reese, C.; Bao, Z., Highly sensitive flexible pressure sensors with microstructured rubber dielectric layers. *Nature Materials* 2010, *9* (10), 859–864.

[5] Gamota, D. R.; Brazis, P.; Kalyanasundaram, K.; Zhang, J., *Printed organic and molecular electronics*. Springer Science & Business Media, New York: 2013.

[6] Rutenberg, I. M.; Scherman, O. A.; Grubbs, R. H.; Jiang, W.; Garfunkel, E.; Bao, Z., Synthesis of polymer dielectric layers for organic thin film transistors via surface-initiated ring-opening metathesis polymerization. *Journal of the American Chemical Society* 2004, *126* (13), 4062–4063.

[7] Muccini, M., A bright future for organic field-effect transistors. *Nature Materials* 2006, *5* (8), 605–613.

[8] Tang, W.; Feng, L.; Yu, P.; Zhao, J.; Guo, X., Highly efficient all-solution-processed low-voltage organic transistor with a micrometer-thick low-k polymer gate dielectric layer. *Advanced Electronic Materials* 2016, *2*, 1500454.

[9] Sze, S. M.; Ng, K. K., *Physics of semiconductor devices*. John Wiley & Sons, Hoboken, New Jersey: 2006.

[10] Peng, X.; Horowitz, G.; Fichou, D.; Garnier, F., All-organic thin-film transistors made of alpha-sexithienyl semiconducting and various polymeric insulating layers. *Applied Physics Letters* 1990, *57* (19), 2013–2015.

[11] Bao, Z.; Feng, Y.; Dodabalapur, A.; Raju, V.; Lovinger, A. J., High-performance plastic transistors fabricated by printing techniques. *Chemistry of Materials* 1997, *9* (6), 1299–1301.

[12] Klauk, H.; Halik, M.; Zschieschang, U.; Schmid, G.; Radlik, W.; Weber, W., High-mobility polymer gate dielectric pentacene thin film transistors. *Journal of Applied Physics* 2002, *92* (9), 5259–5263.

[13] Veres, J.; Ogier, S. D.; Leeming, S. W.; Cupertino, D. C.; Mohialdin Khaffaf, S., Low-k insulators as the choice of dielectrics in organic field-effect transistors. *Advanced Functional Materials* 2003, *13* (3), 199–204.

[14] Park, S. Y.; Park, M.; Lee, H. H., Cooperative polymer gate dielectrics in organic thin-film transistors. *Applied Physics Letters* 2004, *85* (12), 2283–2285.

[15] Huang, W.; Besar, K.; Zhang, Y.; Yang, S.; Wiedman, G.; Liu, Y.; Guo, W.; Song, J.; Hemker, K.; Hristova, K., A high-capacitance salt-free dielectric for self-healable, printable, and flexible organic field-effect transistors and chemical sensor. *Advanced Functional Materials* 2015, *25* (24), 3745–3755.

[16] Jung, S.; Albariqi, M.; Gruntz, G.; Al Hathal, T.; Peinado, A.; Garcia-Caurel, E.; Nicolas, Y.; Toupance, T.; Bonnassieux, Y.; Horowitz, G., A TIPS-TPDO-tetraCN-based n-type

organic field-effect transistor with a cross-linked PMMA polymer gate dielectric. *ACS Applied Materials & Interfaces* 2016, *8* (23), 14701–14708.

[17] Schmidt, G. C.; Höft, D.; Haase, K.; Bellmann, M.; Kheradmand-Boroujeni, B.; Hassinen, T.; Sandberg, H.; Ellinger, F.; Hübler, A. C., Fully printed flexible audio system on the basis of low-voltage polymeric organic field-effect transistors with three layer dielectric. *Journal of Polymer Science Part B: Polymer Physics* 2015, *53* (20), 1409–1415.

[18] Irimia-Vladu, M.; Troshin, P. A.; Reisinger, M.; Shmygleva, L.; Kanbur, Y.; Schwabegger, G.; Bodea, M.; Schwödiauer, R.; Mumyatov, A.; Fergus, J. W., Biocompatible and biodegradable materials for organic field-effect transistors. *Advanced Functional Materials* 2010, *20* (23), 4069–4076.

[19] Liang, L.; Mitsumura, Y.; Nakamura, K.; Uemura, S.; Kamata, T.; Kobayashi, N., Temperature dependence of transfer characteristics of OTFT memory based on DNA-CTMA gate dielectric. *Organic Electronics* 2016, *28*, 294–298.

[20] Wang, C. H.; Hsieh, C. Y.; Hwang, J. C., Flexible organic thin-film transistors with silk fibroin as the gate dielectric. *Advanced Materials* 2011, *23* (14), 1630–1634.

[21] Horowitz, G.; Hajlaoui, R.; Bouchriha, H.; Bourguiga, R.; Hajlaoui, M., The concept of "threshold voltage" in organic field-effect transistors. *Advanced Materials* 1998, *10* (12), 923–927.

[22] Huang, W.; Shi, W.; Han, S.; Yu, J., Hysteresis mechanism and control in pentacene organic field-effect transistors with polymer dielectric. *AIP Advances* 2013, *3* (5), 052122.

[23] Li, H.; Yu, J.; Huang, W.; Shi, W.; Huang, J., High performance pentacene organic field-effect transistors consisting of biocompatible PMMA/silk fibroin bilayer dielectric. *Chinese Physics B 2014, 23* (3), 038505

[24] Ortiz, R. P.; Facchetti, A.; Marks, T. J., High-k organic, inorganic, and hybrid dielectrics for low-voltage organic field-effect transistors. *Chemical Reviews* 2009, *110* (1), 205–239.

[25] Sun, X.; Zhang, L.; Di, C. A.; Wen, Y.; Guo, Y.; Zhao, Y.; Yu, G.; Liu, Y., Morphology optimization for the fabrication of high mobility thin-film transistors. *Advanced Materials* 2011, *23* (28), 3128–3133.

[26] Han, S.; Huang, W.; Shi, W.; Yu, J., Performance improvement of organic field-effect transistor ammonia gas sensor using ZnO/PMMA hybrid as dielectric layer. *Sensors and Actuators B: Chemical* 2014, *203*, 9–16.

[27] Feng, L.; Tang, W.; Zhao, J.; Cui, Q.; Jiang, C.; Guo, X., All-solution-processed low-voltage organic thin-film transistor inverter on plastic substrate. *IEEE Transactions on Electron Devices* 2014, *61* (4), 1175–1180.

[28] Han, S.; Yu, X.; Shi, W.; Zhuang, X.; Yu, J., Solvent-dependent electrical properties improvement of organic field-effect transistor-based on disordered conjugated polymer/insulator blends. *Organic Electronics* 2015, *27*, 160–166.

[29] Li, S.; Feng, L.; Zhao, J.; Guo, X.; Zhang, Q., Low temperature cross-linked, high performance polymer gate dielectrics for solution-processed organic field-effect transistors. *Polymer Chemistry* 2015, *6* (32), 5884–5890.

[30] Shu, L.; Shi, W.; Huang, W.; Yu, J., Performance improvement of pentacene organic field-effect transistor through introducing polymer buffer layers. *Journal of Materials Science: Materials in Electronics* 2014, *25* (12), 5540–5545.

[31] Feng, L.; Tang, W.; Xu, X.; Cui, Q.; Guo, X., Ultralow-voltage solution-processed organic transistors with small gate dielectric capacitance. *IEEE Electron Device Letters* 2013, *34* (1), 129–131.

[32] Yoon, M.-H.; Kim, C.; Facchetti, A.; Marks, T. J., Gate dielectric chemical structure-organic field-effect transistor performance correlations for electron, hole, and ambipolar organic semiconductors. *Journal of the American Chemical Society* 2006, *128* (39), 12851–12869.

[33] Wang, C.; Lee, W.-Y.; Nakajima, R.; Mei, J.; Kim, D. H.; Bao, Z., Thiol–ene cross-linked polymer gate dielectrics for low-voltage organic thin-film transistors. *Chemistry of Materials* 2013, *25* (23), 4806–4812.

[34] Ko, J. M.; Kang, Y. H.; Lee, C.; Cho, S. Y., Electrically and thermally stable gate dielectrics from thiol–ene cross-linked systems for use in organic thin-film transistors. *Journal of Materials Chemistry C* 2013, *1* (18), 3091–3097.

[35] Jang, J.; Nam, S.; Hwang, J.; Park, J.-J.; Im, J.; Park, C. E.; Kim, J. M., Photocurable polymer gate dielectrics for cylindrical organic field-effect transistors with high bending stability. *Journal of Materials Chemistry* 2012, *22* (3), 1054–1060.

[36] Li, S.-X.; Feng, L.-R.; Guo, X.-J.; Zhang, Q., Application of thermal azide–alkyne cycloaddition (TAAC) reaction as a low temperature cross-linking method in polymer gate dielectrics for organic field-effect transistors. *Journal of Materials Chemistry C* 2014, *2* (18), 3517.

[37] Li, S.; Tang, W.; Zhang, W.; Guo, X.; Zhang, Q., Cross-linked polymer-blend gate dielectrics through thermal click chemistry. *Chemistry* 2015, *21* (49), 17762–17768.

[38] Li, S.; Zhang, Q., The dielectric properties of low temperature thermally cross-linked polystyrene and poly(methyl methacrylate) thin films. *RSC Advances* 2015, *5* (37), 28980–28984.

[39] Irimia-Vladu, M.; Sariciftci, N. S.; Bauer, S., Exotic materials for bio-organic electronics. *Journal of Materials Chemistry* 2011, *21* (5), 1350–1361.

[40] Irimia-Vladu, M., "Green" electronics: biodegradable and biocompatible materials and devices for sustainable future. *Chemical Society Reviews* 2014, *43* (2), 588–610.

[41] Shi, W.; Yu, J.; Huang, W.; Zheng, Y., Performance improvement of a pentacene organic field-effect transistor through a DNA interlayer. *Journal of Physics D: Applied Physics* 2014, *47* (20), 205402.

[42] Steckl, A. J., DNA-a new material for photonics? *Nature Photonics* 2007, *1* (1), 3–5.

[43] Stadler, P.; Oppelt, K.; Singh, T. B.; Grote, J. G.; Schwödiauer, R.; Bauer, S.; Piglmayer-Brezina, H.; Bäuerle, D.; Sariciftci, N. S., Organic field-effect transistors and memory elements using deoxyribonucleic acid (DNA) gate dielectric. *Organic Electronics* 2007, *8* (6), 648–654.

[44] Shapiro, R., The prebiotic role of adenine: a critical analysis. *Origins of Life and Evolution of the Biosphere* 1995, *25* (1–3), 83–98.

[45] Irimia-Vladu, M.; Troshin, P. A.; Reisinger, M.; Schwabegger, G.; Ullah, M.; Schwoediauer, R.; Mumyatov, A.; Bodea, M.; Fergus, J. W.; Razumov, V. F., Environmentally sustainable organic field-effect transistors. *Organic Electronics* 2010, *11* (12), 1974–1990.

[46] Irimia-Vladu, M.; Głowacki, E. D.; Troshin, P. A.; Schwabegger, G.; Leonat, L.; Susarova, D. K.; Krystal, O.; Ullah, M.; Kanbur, Y.; Bodea, M. A., Indigo-a natural pigment for high performance ambipolar organic field-effect transistors and circuits. *Advanced Materials* 2012, *24* (3), 375–380.

[47] Marsh, R. E.; Corey, R. B.; Pauling, L., An investigation of the structure of silk fibroin. *Biochimica et Biophysica Acta* 1955, *16*, 1–34.

[48] Sashina, E.; Golubikhin, A. Y.; Novoselov, N.; Tsobkallo, E.; Zaborskii, M.; Goralskii, Y., Study of a possibility of applying the films of the silk fibroin and its mixtures with synthetic polymers for creating the materials of contact lenses. *Russian Journal of Applied Chemistry* 2009, *82* (5), 898–904.

[49] Kim, D.-H.; Kim, Y.-S.; Amsden, J.; Panilaitis, B.; Kaplan, D. L.; Omenetto, F. G.; Zakin, M. R.; Rogers, J. A., Silicon electronics on silk as a path to bioresorbable, implantable devices. *Applied Physics Letters* 2009, *95* (13), 133701.

[50] Parker, S. T.; Domachuk, P.; Amsden, J.; Bressner, J.; Lewis, J. A.; Kaplan, D. L.; Omenetto, F. G., Biocompatible silk printed optical waveguides. *Advanced Materials* 2009, *21* (23), 2411–2415.

[51] Li, X.; Shi, W.; Yu, X.; Yu, J., Performance improvement of organic field-effect transistor-based nitrogen dioxide gas sensor using biocompatible PMMA/silk fibroin bilayer dielectric. *Journal of Materials Science: Materials in Electronics* 2015, *26* (10), 7948–7954.

[52] Singh, B.; Sariciftci, N. S.; Grote, J. G.; Hopkins, F. K., Bio-organic-semiconductor-field-effect-transistor based on deoxyribonucleic acid gate dielectric. *Journal of Applied Physics* 2006, *100* (2), 024514.

[53] Shi, W.; Yu, X.; Zheng, Y.; Yu, J., DNA based chemical sensor for the detection of nitrogen dioxide enabled by organic field-effect transistor. *Sensors and Actuators B: Chemical* 2016, *222*, 1003–1011.

[54] Marinkovic, M.; Belaineh, D.; Wagner, V.; Knipp, D., On the origin of contact resistances of organic thin film transistors. *Advanced Materials* 2012, *24* (29), 4005–4009.

[55] Shi, W.; Yu, J.; Huang, W.; Yu, X.; Zheng, Y., Performance enhancement of poly (3-hexylthiophene) organic field-effect transistor by inserting poly (methylmethacrylate) buffer layer. *Applied Physics Letters* 2013, *102* (11), 111607.

[56] Sekitani, T.; Zschieschang, U.; Klauk, H.; Someya, T., Flexible organic transistors and circuits with extreme bending stability. *Nature Materials* 2010, *9* (12), 1015–1022.

[57] Yu, X.; Zhou, N.; Han, S.; Lin, H.; Buchholz, D. B.; Yu, J.; Chang, R. P.; Marks, T. J.; Facchetti, A., Flexible spray-coated TIPS-pentacene organic thin-film transistors as ammonia gas sensors. *Journal of Materials Chemistry C* 2013, *1* (40), 6532–6535.

[58] Mei, J.; Diao, Y.; Appleton, A. L.; Fang, L.; Bao, Z., Integrated materials design of organic semiconductors for field-effect transistors. *Journal of the American Chemical Society* 2013, *135* (18), 6724–6746.

[59] Briseno, A. L.; Mannsfeld, S. C.; Ling, M. M.; Liu, S.; Tseng, R. J.; Reese, C.; Roberts, M. E.; Yang, Y.; Wudl, F.; Bao, Z., Patterning organic single-crystal transistor arrays. *Nature* 2006, *444* (7121), 913–917.

[60] Shaymurat, T.; Tang, Q.; Tong, Y.; Dong, L.; Liu, Y., Gas dielectric transistor of CuPc single crystalline nanowire for SO2 detection down to sub-ppm levels at room temperature. *Advanced Materials* 2013, *25* (16), 2269–2273.

[61] Torsi, L.; Magliulo, M.; Manoli, K.; Palazzo, G., Organic field-effect transistor sensors: a tutorial review. *Chemical Society Reviews* 2013, *42* (22), 8612–8628.

[62] Yu, J.; Yu, X.; Zhang, L.; Zeng, H., Ammonia gas sensor based on pentacene organic field-effect transistor. *Sensors and Actuators B: Chemical* 2012, *173*, 133–138.

[63] Lai, S.; Demelas, M.; Casula, G.; Cosseddu, P.; Barbaro, M.; Bonfiglio, A., Ultralow voltage, OTFT-based sensor for label-free DNA detection. *Advanced Materials* 2013, *25* (1), 103–107.

[64] Lin, P.; Yan, F., Organic thin-film transistors for chemical and biological sensing. *Advanced Materials* 2012, *24* (1), 34–51.

[65] Wang, B.; Huynh, T. P.; Wu, W.; Hayek, N.; Do, T. T.; Cancilla, J. C.; Torrecilla, J. S.; Nahid, M. M.; Colwell, J. M.; Gazit, O. M., A highly sensitive diketopyrrolopyrrole-based ambipolar transistor for selective detection and discrimination of xylene isomers. *Advanced Materials* 2016, *28* (21), 4012–4018.

[66] Knopfmacher, O.; Hammock, M. L.; Appleton, A. L.; Schwartz, G.; Mei, J.; Lei, T.; Pei, J.; Bao, Z., Highly stable organic polymer field-effect transistor sensor for selective detection in the marine environment. *Nature Communications* 2014, *5*, 2954.

[67] Huang, W.; Yu, J.; Yu, X.; Shi, W., Polymer dielectric layer functionality in organic field-effect transistor based ammonia gas sensor. *Organic Electronics* 2013, *14* (12), 3453–3459.

[68] Zhou, Y.; Han, S.-T.; Huang, L.-B.; Huang, J.; Yan, Y.; Zhou, L.; Roy, V., A low voltage programmable unipolar inverter with a gold nanoparticle monolayer on plastic. *Nanotechnology* 2013, *24* (20), 205202.

[69] Ji, D.; Jiang, L.; Dong, H.; Meng, Q.; Wang, Z.; Zhang, H.; Hu, W., "Double exposure method": a novel photolithographic process to fabricate flexible organic field-effect transistors and circuits. *ACS Applied Materials & Interfaces* 2013, *5* (7), 2316–2319.

[70] Liu, X.; Zhao, H.; Dong, G.; Duan, L.; Li, D.; Wang, L.; Qiu, Y., Multifunctional organic phototransistor-based nonvolatile memory achieved by UV/ozone treatment of the Ta2O5 gate dielectric. *ACS Applied Materials & Interfaces* 2014, *6* (11), 8337–8344.

[71] Sham, M. L.; Li, J.; Ma, P. C.; Kim, J.-K., Cleaning and functionalization of polymer surfaces and nanoscale carbon fillers by UV/ozone treatment: a review. *Journal of Composite Materials* 2009, *43* (14), 1537–1564.

[72] Huang, W.; Yu, X.; Fan, H.; Yu, J., High performance unipolar inverters by utilizing organic field-effect transistors with ultraviolet/ozone treated polystyrene dielectric. *Applied Physics Letters* 2014, *105* (9), 093302.

[73] Guo, X.; Feng, L.; Cui, Q.; Xu, X., Low voltage organic/inorganic hybrid complementary inverter with low temperature all solution processed semiconductor and dielectric layers. *IEEE Electron Device Letters* 2014, *35*, 542–544.

[74] Feng, L.; Cui, Q.; Zhao, J.; Tang, W.; Guo, X., Dual-low-voltage solution processed organic thin-film Transistors with a thick polymer dielectric layer. *IEEE Transactions on Electron Devices* 2014, *61* (6), 2220–2223.

[75] Feng, L.; Sakai, H.; Sakuragawa, Y.; Wang, R.; Zhao, J.; Murata, H.; Guo, X., Stabilization of space charge polarization in ion-dispersed gate dielectric layer of organic transistors by ultraviolet illumination for write-once read-many memory. *Organic Electronics* 2015, *26*, 471–475.

Chapter 2

High-*k* Polymer Nanocomposites for Energy Storage Applications

Asad Mahmood, Abdul Naeem and
Tahira Mahmood

Additional information is available at the end of the chapter

http://dx.doi.org/10.5772/65944

Abstract

High dielectric (high-*k*) polymer nanocomposites that can electrostatically store energy are widely used in electronics and electric power systems due to their high breakdown strengths (E_b), durability, and ability to configure in various shapes. However, these nanocomposites suffer from a limited working temperature regime, thus limiting their extreme applications, such as hybrid and electric vehicles, aerospace power electronics, and deep ground fuel exploration. Furthermore, the E_b and the electric displacement (D) of polymer nanocomposites must be simultaneously enhanced for high-density capacitor applications, which prove to be difficult to modify concurrently. This chapter thoroughly reviews (investigates) the recent developments in the high-*k* polymer nanocomposites synthesis, characterization, and energy storage applications. Consequently, the aim of this chapter is to provide an overview of the novel developmental strategies in order to develop high-dielectric nanocomposites perovskite ceramics that can be incorporated in high-energy-density (HED) applications.

Keywords: polymer nanocomposites, capacitors, perovskite ceramics, dielectric constant, energy density applications

1. Introduction

Highdielectric constant (*high-k*) materials and polymer nanocomposites are under extensive investigation because of their potential applications in organic field-effect transistors (OFETs), invertors, electro-optics, energy, humidity and temperature sensor applications. The magnitude of k for silicon dioxide (SiO_2) is given as 3.9. Materials that exhibit $k > 3.9$ are classified as

INTECH
open science | open minds

© 2017 The Author(s). Licensee InTech. This chapter is distributed under the terms of the Creative Commons Attribution License (http://creativecommons.org/licenses/by/3.0), which permits unrestricted use, distribution, and reproduction in any medium, provided the original work is properly cited.

high-k, and materials whose $k< 3.9$ are classified as low-k materials. Polymer dielectric nanocomposites are generally composed of dielectric polymers as the matrix material, and inorganic/organic fillers as the reinforcement, utilizing the properties of both. Polymers have been found to demonstrate high breakdown strengths along with high energy density (HED), while the fillers, especially dielectric ceramics, have a high dielectric constant (k). The combination of both provides superior dielectric properties depending on the type and nature of the polymer matrices as well as the fillers. The total energy storage density (U) of a capacitor is given as Eq. (1).

$$U = \int \mathbf{E} dD \tag{1}$$

where **E** and D are the applied electric field and electric displacement, respectively. The magnitude of U for the linear dielectrics is defined as Eq. (2).

$$U = 1/2D\mathbf{E} + 1/2k\varepsilon_0\mathbf{E}^2 \tag{2}$$

Where, k is the dielectric constant and $\varepsilon_0 = 8.854 \times 10^{-12}$ F/m is the permittivity of free space. Therefore, the magnitude of U in Eq. (2) depends on the values of both k and **E**. However, limitations exist in ceramic and electrolytic capacitors, such as the applied **E** of a material is restricted to the breakdown strength of the same material. Due to these deficiencies, polymer capacitors have proven to be better alternatives for high energy density applications because of their high E_b, easy and economical processing, and flexibility in designing [1–3].

For high energy density applications, various ceramic fillers have been reported in the recent literature, such as barium titanate ($BaTiO_3$, BT), barium strontium titanate ($Ba_{1-x}Sr_xTiO_3$), barium zirconate titanate ($BaZr_xTi_{1-x}O_3$), lead zirconate titanate ($PbZr_xTi_{1-x}O_3$), metal phenylphosphonates $ATi(O_3PC_6H_5)_3$, where A = Mg, Ca, Sr, Ba, Pb, $MnCoFe_2O_4$, and $MnCuFe_2O_4$ [4–6]. The inorganic-organic nanocomposites have been processed by various techniques. Generally, the presynthesized filler nanoparticles are dispersed in the polymer matrix. Another method known as the *in situ* sol-gel involves the hydrothermal treatment of the filler precursor/polymer system in hydroxide solution. The filler particles are synthesized at the precursor sites in the polymeric system. Despite several modifications in these synthetic procedures, the final materials still exhibit conglomeration, which is responsible for the inconsistent behavior of the system. Furthermore, due to the incompatibility of the polymer-precursor and possible gel formation overtime, the system properties are compromised [7]. Zhu et al. [8] investigated the core-shell structured polymer@BT nanoparticles by using a polymer shells with different elemental properties in order to study the effect of core-shell in detail for surface modifications. A surface-initiated reversible-addition-fragmentation chain transfer (RAFT) polymerization method was used, where the thicknesses of the polymer shells were maintained to the same values. Results of this study indicated that for high energy density applications, high dielectric constant and low electrical conductivity are required. The importance of conductivity of the shell material in the final device performance was also reported in this investigation [8]. Other filler materials that are used include graphene oxide, carbon nanotubes (CNTs), and metal nanoparticles.

This chapter provides a comprehensive understanding of the recent developments in the high-*k* nanocomposites without detailing the complex mathematical models and formulas that have been excessively quoted in various books and reviews. We provide an understanding of the nanocomposites processing, strategies to optimize their dielectric properties, and various filler materials that have been recently used. This section also strives to provide a future perspective in the high-*k* nanocomposites and conclude with some summarized remarks about the topic.

2. Electrical properties

Currently, pure polymers such as biaxially oriented polypropylene (BOPP) are used in energy storage applications, because of their high breakdown strength and low cost. However, pure polymers suffer from low magnitudes of dielectric constant, which results in low performance for energy density applications [2]. To overcome low dielectric constant limitations, high dielectric constant electroceramic nanoparticles are used as filler materials. Generally, due to the small band gap in semiconductors, thermal excitation is responsible for the generation of charge carriers, while the band gap of dielectric materials is relatively large, so electrical contacts or an external sources are responsible for the injection of charge carriers. Transition from one band to another in dielectrics requires high energy [9]. Electrons present in the valence shell of an atom are responsible for the dielectric phenomena, which interacts with externally applied fields, such as an electric or magnetic field. In polar dielectrics, the +ve and –ve charges are responsible for the production of electric dipole, whereas nonpolar dielectric materials lack this inherent dipole at the zero electric field, so an external potential is applied to shift the electron cloud, resulting in dipole phenomena [10]. Despite the use of high-*k* ceramics as filler materials in the development of nanocomposites, another approach is to introduce electrically conductive nanofillers. Such a device configuration has shown superior dielectric response; however, low breakdown strength and high dissipation factor both are disadvantages to be considered [11]. Ceramics became the backbone of the electronic industry after the discovery of ferroelectricity in these materials in 1946. Ferroelectricity was first observed in BT in the mid-1940s and since then, BT has become one of the most essential ferroelectric materials [12]. Before the discovery of BT, steatite, mica, $MgTiO_3$, $CaTiO_3$, and TiO_2 were widely used in capacitors. These materials exhibit $\varepsilon_r \leq 100$, which limits their use in the industry of dielectric capacitors [13–15]. During World War II, the need of high dielectric constant materials increased in capacitor applications [16, 17]. Ferroelectric materials exhibit nonlinear dielectric polarization against an external E due to the presence of permanent dipoles [18]. The polarization in these materials increases with the increasing strength of E, which corresponds to the alignment of dipoles in an applied electric field. The state (P_s) in which the dipoles of the ferroelectrics align with the applied field is called saturation. The polarization decreases with the reduction of E as well as with the reversal of the applied field's direction, however, when the applied electric field is removed ($E = 0$), ferroelectric materials still exhibit some polarization. When the value of the field reaches to a certain point, the materials exhibit zero polarization, which is called coercive field (E_c). Dielectric properties of ferroelectric materials change significantly with temperature near T_c. Generally, magnitude of k increases with an increasing temperature below T_c, reaching a maximum value and then decreases with a further increase in temperature, for example, barium titanate $k \approx 2000$ at room temperature reaches 7000 at T_c(~120°C) [19, 20]. Due to the high dielectric constant of ceramic materials, they are deemed as an efficient filler material for energy storage applications.

An ideal dielectric material resists the flow of charge completely; only allowing displacement of charges, resulting in polarization. In a typical capacitor arraignment, when an alternating E is applied to an ideal dielectric, the current will carry the voltage by a phase angle of $\pi/2$ (i.e., 90°), where no power will be absorbed by the dielectric, and the capacitor will exhibit dielectric loss ($\tan \delta = 0$). On the other hand, the practical materials exhibit $\tan \delta$, and it is because the current-voltage phase angle is not exactly 90°; therefore, the current is slightly lagging behind. The angle and magnitude of the lag are defined as δ and $\tan \delta$, respectively. Dielectric losses are associated with various mechanisms occurring in materials, such as electronic polarization, ion vibration, deformation, and ion migration. Generally, dielectric loss in ceramics is due to ion migration. Temperature and frequency are important factors that influence $\tan \delta$ [21]. Compared to dielectric ceramics, the low magnitudes of k and $\tan \delta$ in polymer dielectrics are attractive for transistor applications and they can be used in the form of thin layers (nanometer layers) as they exhibit low leakage current and high breakdown strength, which contributes to the miniaturization of electrical devices. **Table 1** summarizes the breakdown strength of some common polymers. Pure polymers are not suitable for high energy density applications because of their low dielectric constant.

Polymer	Dielectric strength (V/μm)	Polymer	Dielectric strength (V/μm)
Polyethylene (LD)	200	Polypropylene (biaxially oriented)	200
Polyethylene (HD)	200	Polystyrene	200
Polyethylene (XL)	220	Poly(vinylidene fluoride)	10.2
Polycarbonate	252	Polyester	300
Polyimide	280	Epoxy resin	25–45

Table 1. Dielectric strength of commonly used polymers [22].

High dielectric constants in ceramics originated from ionic polarization, which is associated with the asymmetric migration of the central metal cation in the crystal lattice. In contrast, polymers exhibit various types of polarization, i.e., ionic, electronic, and orientation, that are responsible for their dielectric constants [23, 24]. Typically, depending on the nature of the material and applied field, at least one kind of polarization mechanism is present in dielectrics. Electronic polarization is induced in a dielectric material when subjected to an external electric field, where the electron cloud is displaced relative to the nucleus. This type of polarization may be stimulated in all dielectrics when they are placed under an external electric field. In ionic polarization, the cations and anions in an atom are displaced in opposite directions, resulting in a net dipole moment. The magnitude of the dipole moment generated for each ion pair is equivalent to the product of relative displacement and respective ionic charges. This type of polarization occurs only in ionic materials. Orientation polarization occurs in materials that possess permanent dipole moments, and it is induced when these permanent dipoles align in the direction of an applied electric field. This type of polarization decreases with an increase of temperature [25].

Another polarization characteristic of the multicomponent dielectric systems (such as semi-crystalline polymers, inorganic-organic nanocomposites, and polymer blends) is Maxwell-Wagner-Sillars (MWS) interfacial polarization (IP), which is associated with the reorganization of the interface charges, i.e., electrons and hole accumulated at the interfaces in a heterogeneous system. According to the MWS effect, when an electric field is applied to a heterogeneous dielectric system, the charges accumulate on the surfaces of the different dielectrics exhibiting different retention time. Such a configuration is desirable in achieving a high dielectric constant in the system. However, a drawback to this inhomogeneous distribution of charge in a multicomponent system is that the whole system may fail, therefore, it is important to optimize all the parameters that will define the final properties of the system for better performance [3].

3. Filler materials/polymers

The dielectric properties of polymer nanocomposites are affected by various factors such as the type, size, concentration, and shape of the filler materials and polymer matrix. For high energy density applications, the semiconductive fillers such as titanium dioxide (TiO_2), zinc oxide (ZnO), molybdenum sulfide (MoS_2), and silicon carbide (SiC) have been studied. These fillers have shown considerable promise as candidates for high voltage applications in polymer dielectrics [26]. Among these, TiO_2 presents excellent physiochemical properties and it can be crystallized in the form of rutile, anatase, and brookite structures. Tomara et al. [27] reported the processing of anatase TiO_2-epoxy resin nanocomposites. Various concentrations from 3 to 12% parts per hundred resins per weight were investigated. The broadband dielectric spectroscopy measurements showed five relaxation modes designated as γ, β, intermediate dipolar effect (IDE), α, and interfacial polarization (IP) with varying temperatures and constant frequencies. The γ and β modes were associated with the reorientation of smaller segments and rearrangement of the polarized groups. The IDE, α and IP were attributed to the TiO_2 inclusions, glass to rubber transitions, and electric heterogeneity of the nanocomposites, respectively. The TiO_2-polymer composites were reported to have a superior dielectric response compared to pure polymers [27]. Xie et al. [28] prepared core-shell structured hyperbranched aromatic polyimide grafted $BaTiO_3$ (BT-HBP) hybrid nanofillers. Poly(vinylideneflouride)-trifluoroethylene-chlorofluoethylene(PVDF-TrFE-CFE) was used as the polymer matrix. The dielectric study showed that the 40 vol% of BT-HBP had a high dielectric constant (1485.5 at 1000 Hz) compared to untreated $BaTiO_3$, which was recorded as 206.3 [28].

To further investigate and optimize the dielectric and energy storage properties of polymer nanocomposites, dielectric ceramics with perovskite structures are used as filler materials. **Table 2** presents various ceramic fillers and polymers with respective dielectric constants.

Fillers	*k*	Polymers	*k*
$BaTiO_3$	1700	Nonfluorinated aromatic polyimides	3.2–3.6
PLZT (7/60/40)	2590	Fluorinated polyimide	2.6–2.8
$PbNb_2O_6$	225	Poly(phenyl quinoxaline)	2.8
$SrTiO_3$	2000	Poly(arylene ether oxazole)	2.6–2.8
$CaCu_3Ti_4O_{12}$	~60,000	Poly(arylene ether)	2.9
$La_{1.8}Sr_{0.2}NiO_4$	~100,000	Polyquinoline	2.8
TiO_2	80	Silsesquioxane	2.8–3.0
ZrO_2	25	Poly(norborene)	2.4
SiO_2	3.9	Perfluorocyclobutane polyether	2.4
Al_2O_3	9	Fluorinated poly(arylene ether)	2.7
Ta_2O_5	22	Polynaphthalene	2.2
HfO_2	25	Poly(tetrafluoroethylene)	1.9
$HfSiO_4$	11	Polystyrene	2.6
Y_2O_3	15	Poly(vinylidene fluoride-co-hexafluoropropylene)	~12
La_2O_3	30	Poly(ether ketone ketone)	~3.5

*PLZT, lead lanthanum zirconium titanate.

Table 2. Dielectric constant of commonly used ceramics and polymers for energy storage applications [22].

The problem associated with perovskite ceramics as fillers in polymer matrix is the low breakdown strength of the final system.Due to higher dielectric constant of the ceramic materials than polymers, a significant electric field is concentrated in the polymer matrix compared to the ceramic nanoparticles. This behavior is responsible for the decrease of breakdown strength in the ceramic-polymer nanocomposite systems. To address this issue, dielectric fillers with large aspect ratios, such as (1D) $BaTiO_3$, hexagonal boron nitride nanosheets, and (2D) clay nanosheets are used. These strategies are the tradeoff between the permittivity values and dielectric breakdown. These systems have shown high magnitudes of breakdown strength that may be due to the presence of order traps, an increase in the number of scattering traps to injected charges, and the path tortuosity in the electrical treeing process during breakdown [29]. Some recently used ceramic-polymer composites with their corresponding electrical properties are summarized in **Table 3**.

Fillers	Shape	Polymer	Surface modifier	HOD (J/cm³)	E_b	Refs
$Ba_{0.6}Sr_{0.4}TiO_3$	nf	PVDF	–	6.8	3800 kV/cm	[30]
$Ba_{0.6}Sr_{0.4}TiO_3$	nf	PVDF	Isopropyl dioleic (dioctylphosphate) titanate	6.95	3800 kV/cm	[31]
$Ba_{0.7}Sr_{0.3}TiO_3$	nw	(P(VDF–HFP))	Poly(pentafluorophenyl acrylate)	2.2	200 kV/mm	[32]
$BaTiO_3$	np	PDA	PLA	1.52	120 MV/m	[33]
$BaTiO_3$	np	PVDF	Hydroxylation	3.24	2.25 MV/cm	[34]
$Ba_{1-x}Ca_xTiO_3$	np	PVDF	Hydroxylation	4.72	1.75 MV/cm	[34]
$BaZr_{1-x}Ti_xO_3$	np	PVDF	Hydroxylation	7.74	2.5 MV/cm	[34]
$BaTiO_3$	np	PVDF	SiO_2	6.28	340 MV/m	[35]
$BaTiO_3$	nc	P(VDF-HFP	$NOBF_4$	4.66	150 MV/m	[36]
$BaTiO_3$	nf	PVDF	3-aminopropyltriethoxyilane	5.6	3300 kV/cm	[37]

*nf, nanofibers; np, nanoparticles; nc, nanocubes.

Table 3. Energy density and breakdown strength of the ceramic-polymer nanocomposites recently reported in the literature.

Another complication associated with the nanocomposites is the high dielectric loss. To reduce the dielectric loss, a dielectric layer coated with conductive nanoparticles is artificially synthesized. Such efforts have been reported to be successful, however, a decrease in the dielectric constant was also observed. Moreover, the processing of such systems may cause conductive networks or tunneling effect between the conductive nanoparticles coated on the surface of the dielectric layer, which results in high dielectric loss and high current leakage. One method to overcoming this problem is to distribute the fillers in the polymer matrix homogeneously in order to avoid contact between the filler particles [38]. Another problem that may cause high current leakage, high dielectric loss, low dielectric constant, and low breakdown strength are the compatibility issues between the inorganic and organic nanocomposites. To resolve this problem, "all polymer" approach has been introduced, where a dielectric polymer such as PVDF and its derivatives are used in the form of polymer blends. These attempts have shown superior dielectric properties; however, these polymer blends are known to cause high dielectric loss. To further modify these systems, a "percolative" dielectric composite approach has been developed. In this method, a conductive organic domain is dispersed in a dielectric polymer matrix. The space charge accumulates at the interface of the conductive organic and dielectric domains. A high dielectric response is achieved because the conductive domain acts as the "super dipole", resulting in high polarization. The magnitude of the dielectric constant reaches a maximum, with an increasing volume fraction of the organic domain, up to just below the percolation threshold. All polymer systems have the advantage over inorganic-organic polymer composites, such as easy processing, economical, and light weight; however, such systems also suffer from high leakage current and low breakdown strength [39].

The extensive studies that have been reported for various filler materials in the polymer matrix only highlight the advantages of the final systems. It is important to critically analyze the underlining physics and chemistry of the filler materials, polymer matrix, and the dielectric response of the multicomponent systems. An essential aspect of the reported results is the reproducibility for practical device performance.

4. Interfacial chemistry

The dielectric constant of the polymer nanocomposites can be increased by loading a high concentration of the dielectric constant fillers. However, high constant, i.e., BT (60%) leads to agglomeration of the filler particles in the polymer matrix, which results in pore formation and subsequently low dielectric constant and high dielectric loss. In order to homogeneously load high concentration of the dielectric fillers, the interfacial chemistry between the filler particles and the polymer matrix must be optimized. Furthermore, the intergranular interaction should also be investigated for homogeneous distribution and conglomerates formed in the polymer matrix. Generally, the interfacial polarization is considered as the primary polarization mechanism in the polymer nanocomposites. The energy density capacity of the polymer nanocomposites can be increased by modifying the interfacial chemistry between the filler materials. By applying this technique, high magnitudes of dielectric constant and breakdown strength can be achieved. Therefore, for better performance of the device, surface interaction between the filler materials itself and the filler-polymer matrix is important [29]. To do so, it is vital to understand the affinity between the polymer matrix and filler particles.

One strategy for addressing this issue is modifying the surface of the filler materials. The surface modifier consists of two major components: the first one is the functional group that attaches the modifier to the filler particle's surface such as, $-PO_4^{3-}$, $-OH$, $-SO_3H^-$, $-SO_3^{2-}$, $-COOH$, $-NR_3^{z+}$, and $-Coo^-$; and the second component is the macromolecular chain, solvable and dispersible in different media such as polyether, polyester, polyolefin, and polyacrylate. For example, phosphoric acid and silane coupling agents are used for surface modification of the BT nanoparticles. Moreover, the molecular structure of the surface modifier must be given considerable attention, making sure it is similar to the polymer matrix; otherwise it will result in the formation of pores, voids, and cracks in the polymer nanocomposites. The problem of high leakage current and dielectric loss associated with the silane coupling agent is due to the unabsorbed residues species in the final system [40, 41]. Niu et al. [42] studied the relationship between the surface modifier structure and BT-PVDF nanocomposites for energy storage applications. The schematic surface modification, composites formation, and energy density graph are illustrated in **Figure 1**.

They reported the use of four types of modifiers which belong to the carboxylic acids, i.e., 2, 3, 4, 5-tetrafluorobenzoic acid, 4-(trifluoromethyl) phthalic acid, tetrafluorophthalic acid, and

http://dx.doi.org/10.5772/65944

phthalic acid designated as F4C, F3C2, F4C2, and C2, respectively. The breakdown strength was observed to improve in all four cases. The energy density of the BT/PVDF nanocomposites with F4C and C2 was increased by 35.7 and 37.7%, respectively [42]. Zhou et al. [43] studied the surface hydroxylated BT (h-BT)/PVDF nanocomposites, where crude BT (c-BT) nanoparticles were transferred to h-BT nanoparticles using the H_2O_2 aqueous solution. Results showed that h-BT presented high thermal stability, low frequency dependent dielectric response, low dielectric loss, and high magnitude of dielectric constant compared to c-BT-based nanocomposites [43]. Mukherjee et al. [44] reported a surface modification of the BT nanoparticles by using the plasma enhanced chemical vapor deposition (CVD) method, where the reactive amine groups were implemented as surface modification agents. The final modified nanoparticles were treated with epoxide monomers to form nanocomposites. A comparison between the modified BT/epoxide and the unmodified BT/epoxide nanocomposites showed that the former exhibited superior thermal and electrical properties to the latter [44].

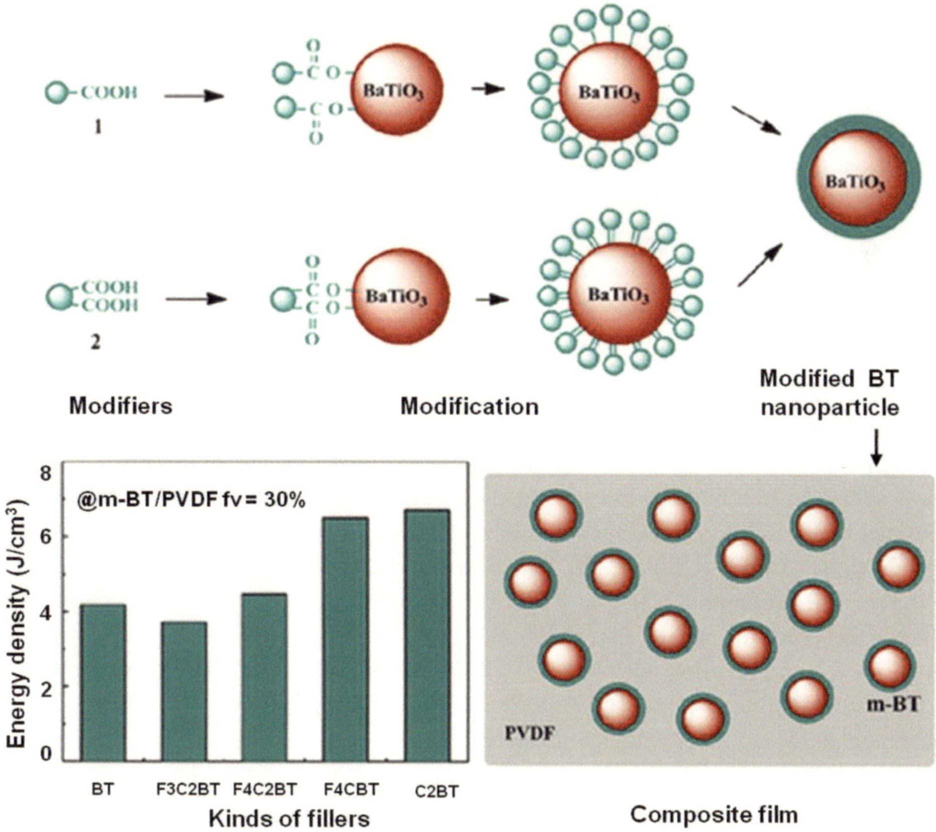

Figure 1. Scheme process of nanoparticle modification and incorporation into the polymer matrix and discharged energy density of the nanocomposites filled with BT and m-BT nanoparticles [42].

Surface modifications of the filler material are important as it decreases the probability of contact between the filler materials, and it improves the compatibility between the fillers and the polymer matrix. Through this procedure, the properties of the polymer nanocomposites can be enhanced for energy storage applications.

5. Processing strategies

The dielectric and energy storage properties of the polymer nanocomposites can also be tuned through the preparation methods. The processing techniques not only include the composite formation, but also the synthesis of the filler materials in various sizes and shapes. This section only summarizes the methods for the formation of filler-polymer nanocomposites. The most adaptable synthesis approach for the processing of the polymer composites in the recent literature is liquid-phase assisted dispersion. In this method, the filler materials and the polymer are mixed in a desired solvent, followed by stirring for a long duration, i.e., 24 h for homogeneous mixing. The mixture is then transferred into a glass substrate, where the thickness of the film is maintained. The film is further dried at various desired temperatures and times in a vacuum/air atmosphere to remove the solvent, then annealed at desired temperatures, i.e., BT@TiO_2-nanofibers-P (VDF-HFP) at 200°C/7 min and subsequently quenched in an ice bath [45]. Another approach is the solution casting method, where a polymer is dissolved in a suitable solvent and a predetermined weight-percent of the filler material is added to the solution. The solution is thoroughly mixed using an ultrasonic bath or a magnetic stirrer, and then spin coated on a glass substrate. The process is repeated until the desired thickness is achieved, then the film is annealed at some low temperature (~80°C) for crystallization, and the solvent is removed by thermal dehydration in an oven.

Xu et al. [46] processed a three-phase system using polyimide (PI), BT, and multiwall carbon nanotubes (MWCNTs) by the electrospinning technique. MWCNTs were homogeneously dispersed in the PI matrix and subsequently collected by using the BT particles. The solution was subjected to electrospinning using 20 kV/20 cm from the spinneret to the collector, keeping the flow rate of 1.2 mL/h at room temperature. The resultant nanofibers were treated with heat, at 70°C/6h (vacuum), 250°C/1 h in N_2. The composite nanofibers were further treated at 5 MPa pressure at 300°C in a vacuum for 5 min using hot-pressing techniques. A high dielectric constant of 1061.98 (at 100 Hz) and an energy density of 4.773 J/cm^3 were recorded for the composition with 40 vol% BTNPs/10 vol% MWCNTs [46]. Zhang et al. [47] reported a novel strategy represented in **Figure 2** for the processing of the BT and PVDF nanocomposites. Nanofibers were synthesized using electrospinning, and the surfaces were modified using 1H,1H,2H,2H-perfluorooctyl trimethoxysilane. Finally, the modified BT and PVDF nanocomposites were processed using solution blending, where the dielectric properties of the final composites were improved. In addition, this technique has an advantage over the grafting method for processing the nanocomposites due to its low cost and simple procedure [47].

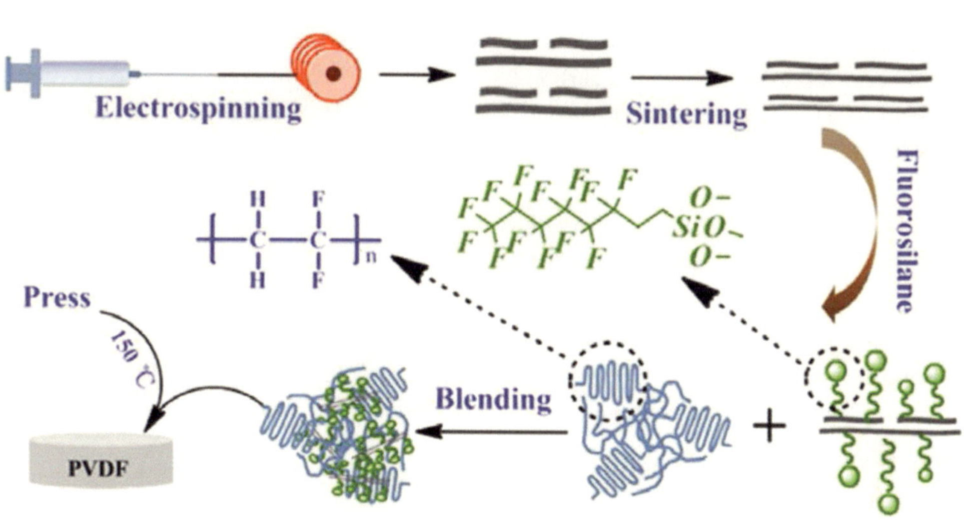

Figure 2. Schematic illustration of preparation of the nanocomposites of PVDF and fluorosilane-modified BT nanofibers [47].

6. Future trends

To further enhance the dielectric response and physiochemical stability of the high-*k* dielectric nanocomposites is the fabrication of high aspect ratio filler materials. Tang et al. [48] studied the effect of different aspect ratios of the BT nanowires on the dielectric properties of the BT/PVDF nanocomposites. The aspect ratios of the nanowires were tuned by controlling the temperature during hydrothermal synthesis, where they were varied from 9.3 to 45.8 corresponding to the temperature ranges from 150 to 240°C, respectively. A direct relationship was observed with the aspect ratio of nanocomposites and dielectric constant. A high dielectric constant of 44.3 was reported for the BT-30 vol% nanocomposites with an aspect ratio of 45.8, which was 30.7% higher than BT samples with an aspect ratio of 9.3 while 352% higher than polymer matrix samples [48]. Despite the high dielectric constant fillers for high dielectric nanocomposite applications, such as BT and $PbZrTiO_3$-based derivatives, these filler materials still suffer from some disadvantages, which include low dielectric response even with high loading concentration, low breakdown strength, and deteriorated flexibility. Furthermore, lead-based compounds are toxic in nature and are expected to be banned in the near future. An alternative to ceramic fillers is the graphene-based nanocomposites for high-energy and ultracapacitor applications. Graphene is a one atom thick sheet of carbon, in which carbon atoms exhibit sp^2 hybridization which correspond to excellent electron mobility, mechanothermal stability, and high flexibility. Graphene oxide (GO) can be easily dispersed in the aqueous solution and the electrical properties can be optimized by using various thermal and chemical approaches such as partial reduction. Thus, the GO/polymer nanocomposites can prove an efficient system for high dielectric applications [49, 50]. Recently, attention has been given to the carbon nanotubes (CNTs), a crystalline form of carbon because of their unique physical and mechanical properties. CNTs are considered to be potential candidates for the formation of nanocomposites, which can improve their electrical, thermal, and mechanical properties [51]. Although extensive studies have reported the dielectric properties of the nanocomposites, very few investigate the dielectric properties vs. temperature [52]. These properties must be thoroughly investigated under mechanical and dielectric thermal stability for practical device applications. Recent advances in the computational material science, physics, and chemistry have developed sophisticated algorithms for predicting the physical and chemical properties of the system. These methods and molecular dynamic techniques must be utilized to design a more reliable and comprehensive system for energy storage applications.

7. Conclusions

For high-energy-density applications, materials should exhibit the high-energy storage capacity. For this purpose, materials should demonstrate high dielectric permittivity, low leakage current, low dielectric loss, and high breakdown strength. All these parameters are crucial for implementing laboratorial research in industrial processes. Furthermore, an economical device fabrication strategy should be adopted for practical and commercial applications; however, a single material does not ideally express all these properties as one system; therefore, a tradeoff is used for device performance between the material properties for different applications. From previous discussion, it has been established that the energy storage capability of the dielectric nanocomposites is strongly influenced by the choice of filler materials, shape and size, morphology, processing methods, use of surfactants or surface modifying agents, and polymer matrix. For better storage applications, all these parameters must be optimized and new strategies should be developed for better processing and understanding of the device fabrications.

Author details

Asad Mahmood*, Abdul Naeem and Tahira Mahmood

*Address all correspondence to: naeem@upesh.edu.pk

National Centre of Excellence in Physical Chemistry, University of Peshawar, Peshawar, Pakistan

http://dx.doi.org/10.5772/65944

References

[1] X. Huang, P. Jiang, Core-shell structured high-k polymer nanocomposites for energy storage and dielectric applications, Advanced Materials, 27 (2015) 546–554.

[2] Q. Li, K. Han, M.R. Gadinski, G. Zhang, Q. Wang, High energy and power density capacitors from solution-processed ternary ferroelectric polymer nanocomposites, Advanced Materials, 26 (2014) 6244–6249.

[3] J. Yuan, S. Yao, P. Poulin, Dielectric Constant of Polymer Composites and the Routes to High-k or Low-k Nanocomposite Materials, Polymer Nanocomposites: Electrical and Thermal Properties, Springer, Switzerland, 2016.

[4] S. Webb, I. Ian Jackson, J.F. Gerald, Viscoelasticity of the titanate perovskites $CaTiO_3$ and $SrTiO_3$ at high temperature, Physics of the Earth and Planetary Interiors, 115 (1999) 259–291.

[5] V. Mathivanan, M. Haris, Studies on solution-grown pure and doped sodium potassium tartrate crystals, Spectrochimica Acta. Part A, Molecular and Biomolecular Spectroscopy, 102 (2013) 341–349.

[6] C. Eckart, Principle of Underwater Sound, (Ed.), Wakefield, Massachusetts, 1946.

[7] M.D. Toomey, K. Gao, G.P. Mendis, E.B. Slamovich, J.A. Howarter, Hydrothermal synthesis and processing of barium titanate nanoparticles embedded in polymer films, ACS Applied Materials & Interfaces, 7 (2015) 28640–28646.

[8] M. Zhu, X. Huang, K. Yang, X. Zhai, J. Zhang, J. He, P. Jiang, Energy storage in ferroelectric polymer nanocomposites filled with core-shell structured polymer@$BaTiO_3$ nanoparticles: understanding the role of polymer shells in the interfacial regions, ACS Applied Materials & Interfaces, 6 (2014) 19644–19654.

[9] G.N. Gerasimov, L.I. Trakhtenberg, Chapter 10 Vapor Deposited Composite Films Consisting of Dielectric Matrix with Metal/Semiconductor Nanoparticles, 1st ed., Elsevier, USA, 2007.

[10] K. Heileman, J. Daoud, M. Tabrizian, Dielectric spectroscopy as a viable biosensing tool for cell and tissue characterization and analysis, Biosensors & Bioelectronics, 49 (2013) 348–359.

[11] K. Yang, X. Huang, L. Xie, C. Wu, P. Jiang, T. Tanaka, Core-shell structured polystyrene/$BaTiO_3$ hybrid nanodielectrics prepared by in situ RAFT polymerization: a route to high dielectric constant and low loss materials with weak frequency dependence, Macromolecular Rapid Communications, 33 (2012) 1921–1926.

[12] A. Mahmood, A. Naeem, Y. Iqbal, T. Mahmood, A. Ullah, Electrical characterization of Mn doped-$(Ba_{0.3}Sr_{0.7})Mnx(Ti_{0.9}Zr_{0.1})_{1-x}O_3$ ceramics, Materials Research Bulletin, 72 (2015) 13–19.

[13] I. Apostol, K.V. Saravanan, C.J.A. Monty, P.M. Vilarinho, Solar physical vapor deposition: a new approach for preparing magnesium titanate nanopowders, Applied Surface Science, 285 (2013) 49–55.

[14] A. Ullah, Y. Iqbal, T. Mahmood, A. Mahmood, A. Naeem, M. Hamayun, Kinetic analysis on the synthesis of $Mg_{0.95}Zn_{0.05}TiO_3$ microwave dielectric ceramic by polymeric precursor method, Ceramics International, 41 (2015) 15089–15096.

[15] E.E.C. Oliveira, A.G. D'Assuncao, J.B.L. Oliveira, A.M. Cabral, Small size dual-band rectangular dielectric resonator antenna based on calcium titanate ($CaTiO_3$), Microwave and Optical Technology Letters, 54 (2012) 976–979.

[16] I.S. Seo, W.S. Chin, D.G. Lee, Characterization of electromagnetic properties of polymeric composite materials with free space method, Composite Structures, 66 (2004) 533–542.

[17] S.M. Abbas, M. Chandra, A. Verma, R. Chatterjee, T.C. Goel, Complex permittivity and microwave absorption properties of a composite dielectric absorber, Composites Part A: Applied Science and Manufacturing, 37 (2006) 2148–2154.

[18] Q.-D. Ling, D.-J. Liaw, C. Zhu, D.S.-H. Chan, E.-T. Kang, K.-G. Neoh, Polymer electronic memories: materials, devices and mechanisms, Progress in Polymer Science, 33 (2008) 917–978.

[19] G. Schileo, Recent developments in ceramic multiferroic composites based on core/shell and other heterostructures obtained by sol-gel routes, Progress in Solid State Chemistry, 41 (2013) 87–98.

[20] E. Sun, W. Cao, Relaxor-based ferroelectric single crystals: growth, domain engineering, characterization and applications, Progress in Materials Science, 65 (2014) 124–210.

[21] L.B. Kong, T.S.Z. S. Li, J.W. Zhai, F.Y.C. Boey, J. Ma, Electrically tunable dielectric materials and strategies to improve their performances, Progress in Materials Science, 55 (2010) 840–893.

[22] P. Barber, S. Balasubramanian, Y. Anguchamy, S. Gong, A. Wibowo, H. Gao, H.J. Ploehn, Z.L. Hans-Conrad, Polymer composite and nanocomposite dielectric materials for pulse power energy storage, Materials, 2 (2009) 1697–1733.

[23] B.C. Riggs, S. Adireddy, C.H. Rehm, V.S. Puli, R. Elupula, D.B. Chrisey, Polymer nanocomposites for energy storage applications, Materials Today: Proceedings, 2 (2015) 3853–3863.

http://dx.doi.org/10.5772/65944

[24] A. Naeem, A. Mahmood, Y. Iqbal, A. Ullah, T. Mahmood, M. Humayun, Dielectric and impedance spectroscopic studies on $(Ba_{0.5}Sr_{0.5})Mn_x(Ti_{0.95}Fe_{0.05})_{1-x}O_3$ ceramics synthesized by using sol–gel method, Journal of Alloys and Compounds, 645 (2015) 290–296.

[25] Z.-G. Ye, Handbook of Advanced Dielectric, Piezoelectric and Ferroelectric Materials: Synthesis, Properties and Applications. Woodhead Publishing Limited, England. Cambridge, 2008.

[26] Q. Jia, X. Huang, G. Wang, J. Diao, P. Jiang, MoS_2 nanosheet superstructures based polymer composites for high-dielectric and electrical energy storage applications, The Journal of Physical Chemistry C, 120 (2016) 10206–10214.

[27] G.N. Tomara, A.P. Kerasidou, A.C. Patsidis, P.K. Karahaliou, G.C. Psarras, S.N. Georga, C.A. Krontiras, Dielectric response and energy storage efficiency of low content TiO_2-polymer matrix nanocomposites, Composites Part A: Applied Science and Manufacturing, 71 (2015) 204–211.

[28] L. Xie, X. Huang, Y. Huang, K. Yang, P. Jiang, Core-shell structured hyperbranched aromatic polyamide/$BaTiO_3$ hybrid filler for poly(vinylidene fluoride-trifluoroethylene-chlorofluoroethylene) nanocomposites with the dielectric constant comparable to that of percolative composites, ACS Applied Materials & Interfaces, 5 (2013) 1747–1756.

[29] X. Zhang, Y. Shen, Q. Zhang, L. Gu, Y. Hu, J. Du, Y. Lin, C.W. Nan, Ultrahigh energy density of polymer nanocomposites containing $BaTiO_3$@TiO_2 nanofibers by atomic-scale interface engineering, Advanced Materials, 27 (2015) 819–824.

[30] S. Liu, S. Xue, W. Zhang, J. Zhai, G. Chen, The influence of crystalline transformation of $Ba_{0.6}Sr_{0.4}TiO_3$ nanofibers/poly(vinylidene fluoride) composites on the energy storage properties by quenched technique, Ceramics International, 41 (2015) S430–S434.

[31] Z.B. Pan, L.M. Yao, J.W. Zhai, S.H. Liu, K. Yang, H.T. Wang, J.H. Liu, Fast discharge and high energy density of nanocomposite capacitors using $Ba_{0.6}Sr_{0.4}TiO_3$ nanofibers, Ceramics International, 42 (2016) 14667–14674.

[32] S. Wang, X. Huang, G. Wang, Y. Wang, J. He, P. Jiang, Increasing the energy efficiency and breakdown strength of high-energy-density polymer nanocomposites by engineering the $Ba_{0.7}Sr_{0.3}TiO_3$ nanowire surface via reversible addition-fragmentation chain transfer polymerization, The Journal of Physical Chemistry C, 119 (2015) 25307–25318.

[33] Y. Fan, X. Huang, G. Wang, P. Jiang, Core-shell structured biopolymer@$BaTiO_3$ nanoparticles for biopolymer nanocomposites with significantly enhanced dielectric properties and energy storage capability, The Journal of Physical Chemistry C, 119 (2015) 27330–27339.

[34] S. Adireddy, V.S. Puli, T.J. Lou, R. Elupula, S.C. Sklare, B.C. Riggs, D.B. Chrisey, Polymer-ceramic nanocomposites for high energy density applications, Journal of Sol-Gel Science and Technology, 73 (2014) 641–646.

[35] K. Yu, Y. Niu, Y. Bai, Y. Zhou, H. Wang, Poly(vinylidene fluoride) polymer based nanocomposites with significantly reduced energy loss by filling with core-shell structured $BaTiO_3/SiO_2$ nanoparticles, Applied Physics Letters, 102 (2013) 102903.

[36] S.S. Parizi, A. Mellinger, G. Caruntu, Ferroelectric barium titanate nanocubes as capacitive building blocks for energy storage applications, ACS Applied Materials & Interfaces, 6 (2014) 17506–17517.

[37] S. Liu, S. Xue, W. Zhang, J. Zhai, Enhanced dielectric and energy storage density induced by surface-modified $BaTiO_3$ nanofibers in poly(vinylidene fluoride) nanocomposites, Ceramics International, 40 (2014) 15633–15640.

[38] S. Luo, S. Yu, R. Sun, C.P. Wong, Nano Ag-deposited $BaTiO_3$ hybrid particles as fillers for polymeric dielectric composites: toward high dielectric constant and suppressed loss, ACS Applied Materials & Interfaces, 6 (2014) 176–182.

[39] M.S. Islam, Y. Qiao, C. Tang, H.J. Ploehn, Terthiophene-containing copolymers and homopolymer blends as high-performance dielectric materials, ACS Applied Materials & Interfaces, 7 (2015) 1967–1977.

[40] N. Kamezawa, D. Nagao, H. Ishii, M. Konno, Transparent, highly dielectric poly(vinylidene fluoride) nanocomposite film homogeneously incorporating $BaTiO_3$ nanoparticles with fluoroalkylsilane surface modifier, European Polymer Journal, 66 (2015) 528–532.

[41] H. Luo, D. Zhang, C. Jiang, X. Yuan, C. Chen, K. Zhou, Improved dielectric properties and energy storage density of poly(vinylidene fluoride-co-hexafluoropropylene) nanocomposite with hydantoin epoxy resin coated $BaTiO_3$, ACS Applied Materials & Interfaces, 7 (2015) 8061–8069.

[42] Y. Niu, Y. Bai, K. Yu, Y. Wang, F. Xiang, H. Wang, Effect of the modifier structure on the performance of barium titanate/poly(vinylidene fluoride) nanocomposites for energy storage applications, ACS Applied Materials & Interfaces, 7 (2015) 24168–24176.

[43] T. Zhou, J.W. Zha, R.Y. Cui, B.H. Fan, J.K. Yuan, Z.M. Dang, Improving dielectric properties of $BaTiO_3$/ferroelectric polymer composites by employing surface hydroxylated $BaTiO_3$ nanoparticles, ACS Applied Materials & Interfaces, 3 (2011) 2184–2188.

[44] N. Mukherjee, D. Wavhal, R.B. Timmons, Composites of plasma surface functionalized barium titanate nanoparticles covalently attached to epoxide matrices: synthesis and evaluation, ACS Applied Materials & Interfaces, 2 (2010) 397–407.

[45] X. Zhang, Y. Shen, B. Xu, Q. Zhang, L. Gu, J. Jiang, J. Ma, Y. Lin, C.W. Nan, Giant energy density and improved discharge efficiency of solution-processed polymer nanocomposites for dielectric energy storage, Advanced Materials, 28 (2016) 2055–2061.

[46] W. Xu, Y. Ding, S. Jiang, L. Chen, X. Liao, H. Hou, Polyimide/$BaTiO_3$/MWCNTs three-phase nanocomposites fabricated by electrospinning with enhanced dielectric properties, Materials Letters, 135 (2014) 158–161.

[47] X. Zhang, Y. Ma, C. Zhao, W. Yang, High dielectric constant and low dielectric loss hybrid nanocomposites fabricated with ferroelectric polymer matrix and $BaTiO_3$ nanofibers modified with perfluoroalkylsilane, Applied Surface Science, 305 (2014) 531–538.

[48] H. Tang, Z. Zhou, H.A. Sodano, Relationship between $BaTiO_3$ nanowire aspect ratio and the dielectric permittivity of nanocomposites, ACS Applied Materials & Interfaces, 6 (2014) 5450–5455.

[49] S. Cho, J.S. Lee, J. Jang, Poly(vinylidene fluoride)/NH_2-treated graphene nanodot/reduced graphene oxide nanocomposites with enhanced dielectric performance for ultrahigh energy density capacitor, ACS Applied Materials & Interfaces, 7 (2015) 9668–9681.

[50] M. Yasin, T. Tauqeer, S.M.H. Zaidi, S.E. San, A. Mahmood, M.E. Köse, B. Canimkurbey, M. Okutan, Synthesis and electrical characterization of graphene oxide films, Thin Solid Films, 590 (2015) 118–123.

[51] C.-C. Wu, C.-C. Su, C.-F. Yang, Preparation, structure and properties of carbon nanotube reinforced polymer nanocomposites, Synthetic Metals, 205 (2015) 98–105.

[52] M. Rajib, M. Arif Ishtiaque Shuvo, H. Karim, D. Delfin, S. Afrin, Y. Lin, Temperature influence on dielectric energy storage of nanocomposites, Ceramics International, 41 (2015) 1807–1813.

Chapter 3

Electrical Properties of Different Polymeric Materials and their Applications: The Influence of Electric Field

Sk Manirul Haque, Jorge Alfredo Ardila Rey,
Abdullahi Abubakar Masúd, Yunusa Umar and
Ricardo Albarracin

Additional information is available at the end of the chapter

http://dx.doi.org/10.5772/67091

Abstract

In this chapter, a comprehensive study on the general structure of polymers, their properties and applications has been carried out. In particular, the application of polymers for insulating high-voltage (HV) equipment has been reported, together with the effect of electric field when they are subjected to HV stress. Experimental results related to the effect of partial discharge (PD) on polymeric insulations have been reported and discussed. Practical implications of the results have been discussed, and recommendations are made for future improvement. It is important to obtain new information regarding novel polymeric materials such as nano-polymers that can possibly outperform the currently used ones. It is also vital to investigate the right information for electrical equipment, i.e. by using the appropriate polymer as solid insulation, minimizing the presence of any metallic sharp object and any other conducting path during manufacture in order to avoid any type of internal or external PD.

Keywords: partial discharge, high-voltage, cross-linked polyethylene, polymeric insulation, electric field

1. Introduction

For the last many decades, tremendous increase in the usage of polymers and polymer-based materials can be seen for various applications. One of the important areas is the usage of polymers for electrical applications. Research studies have been focused on the development of polymers with low and high dielectric constant. Polymers having high dielectric constant can be used for applications such as film capacitors, artificial muscle, while the ones with

INTECH
open science | open minds

© 2017 The Author(s). Licensee InTech. This chapter is distributed under the terms of the Creative Commons Attribution License (http://creativecommons.org/licenses/by/3.0), which permits unrestricted use, distribution, and reproduction in any medium, provided the original work is properly cited. (cc) BY low dielectric constants are used in the field of capacitors and dielectric materials. Thus, it is important to study the dispersion and compatibility of these fillers with polymers since the inter-facial adhesion will reflect in their performance. This is achieved by studying the electric properties, such as the dielectric constant, loss factor and dissipation factor as functions of temperature and frequency. One major challenge is the preparation of a polymer having good adhesion to substrate, good thermal stability, high glass transition temperature and low moisture absorption suitable as a dielectric material for a high-voltage insulation system. Incorporation of voids into polymers will help to decrease the dielectric properties of polymers. Size and distribution of voids in polymers are very important in the performance of the polymers. Internal (void) and corona partial discharge (PD) are the most common defects that affect the polymeric insulations used for high-voltage insulation system. It is well known that discharge impacting at the void surface deposit conducting layers (e.g. oxalic acid and water) around the void boundary is due to chemical decomposition reactions of the polymer. These conducting charge carriers get trapped on the surface and may change the PD mechanism due to variations in the boundary conditions. In a void, streamer discharges prevail over the first hour of stressing and afterwards these normally change to Townsend discharges as a result of the change in resistivity of the surface due to sustained charge carriers.

This chapter analyses the electrical properties of different polymeric materials and their applications and in particular, investigates the properties of the most commonly used polymers for high-voltage applications. It is intended to present experimental results of the discharge development of voids in polymeric materials over a long high-voltage stressing periods—when the initial deterioration of the insulation material starts developing. The PD patterns and degradation over different void dimensions and arrangements will be reported in order to determine the pattern variations due to the electric field and their effect on the chemistry and composition of the material.

2. Polymer chemistry

A polymer is an organic compound with natural or synthetic preparation and has a high molecular weight due to repetitive structural units. Materials that are not conducting under normal conditions have the ability to store charge, known as dielectrics. The properties of dielectrics of a material in an electrical field are now a wide area of interest for the researchers of physics, chemistry, biology, materials science as well as for electrical engineering. Organic polymers are stable at high temperatures, have the ability to resist radiation, high mechanical, electrical stresses, chemical attacks from the extreme environments [1].

Polymeric materials have a low or high dielectric constant, and it is directly related to the permittivity, which is the ability of materials to polarize under the electrical field. It can be calculated by the permittivity of the dielectric to the permittivity of the vacuum. In a simple way, when the dielectric constant is more, polarization will be more in the electric field. Basically, the dielectric materials are made of inorganic compounds like mica and silicon dioxide. In the early 1960s, the polymers like aromatic [1, 2] and polyvinyl fluoride [2] are used as dielectric materials for capacitors. The presence of highly polarizable groups like aromatic rings, bromine and iodine will increase the dielectric constant of materials. In recent times, high dielectric materials (**Table 1**) and low dielectric materials (**Table 2**) are used for capacitors.

http://dx.doi.org/10.5772/67091

Materials	Dielectric permittivity
$BaTiO_3$	1700
PMN-PT (65/35)	3640
$PbNb_2O_6$	225
PLZT (7/60/40)	2590
SiO_2	3.9
Al_2O_3	9
Ta_2O_5	22
TiO_2	80
$SrTiO_3$	2000
ZrO_2	25
HfO_2	25
$HfSiO_4$	11
La_2O_3	30
Y_2O_3	15
α-LaAlO3	30
$CaCu_3Ti_4O_{12}$	60,000
$La_{1.8}Sr_{0.2}NiO_4$	100,000

Note: PMN-PT (65/35) for 65% lead magnesium niobate and 35% lead titanate. PLZT (7/60/40) for lead lanthanum zirconium titanate.

Table 1. High dielectric materials for capacitors [3].

Materials	Dielectric permittivity
Non-fluorinated aromatic polyimides	3.2–3.6
Fluorinated polyimide	2.6–2.8
Poly(phenyl quinoxaline)	2.8
Poly(arylene ether oxazole)	2.6–2.8
Polyquinoline	2.8
Silsesquioxane	2.8–3.0
Poly(norborene)	2.4
Perfluorocyclobutane polyether	2.4
Fluorinated poly(arylene ether)	2.7
Polynaphthalene	2.2
Poly(tetrafluoroethylene)	1.9
Polystyrene	2.6
Poly(vinylidene fluoride-co-hexafluoropropylene)	12
Poly(ether ketone ketone)	3.5

Table 2. Low dielectric materials for capacitors [3].

2.1. Aromatic dielectric materials

2.1.1. Polyimides

Aromatic polyimides have excellent dielectric properties and are very useful materials at high temperatures and voltage. These materials are chosen as inter-layer dielectrics in thin film microelectronics packaging because it is based on a mature technology and satisfy the requisite properties to maintain the mechanical stresses associated with microelectronic fabrication processes. These also include high thermal stability with glass transition temperature, Tg > 300°C, low dielectric constant, easy processing and good chemical stability. Polyimides are a class of polymers, which are synthesized by using two monomers, a dianhydride and a diamine. The soluble polyamic acid (PAA) intermediates undergo a thermal imidization with the evolution of water to form the insoluble polyimide. The undesirable properties of polyimides, large amounts of moisture absorption and high dielectric constant are eliminated by developing a different polyimide. The polyimide properties can be improved by introducing silicone, a fluorinated flexible bridge monomer with the diamine, dianhydride monomers. The new developed method mingles the voids of air by making nano-foams in the film structure. It is necessary to avoid the voids because this factor is mainly responsible for lowering the dielectric constant of the polyimide-based material, and it is only possible when the dielectric constant is equal to 1.

The formation of polyimides is based upon the PAA [4, 5], and it is formed from pyromellitic dianhydride (PMDA) and oxydianiline (ODA). The aromatic polyimides, have a linear, symmetric dimide structure that provides itself a small effective dipole moment for the repeat units and contributes to its low dielectric constant. The polyimide has an ability to absorb moisture that will escalate the polyimide layers, which will significantly increase the dielectric constant.

The general process to reduce the dielectric constant of polyimides is the inclusion of organofluorine components, in the form of pendant perfluoroalkyl groups. The polyamic acid precursor for fluorinated polyimides is commonly based on hexafluorodianhydrideoxydianiline (HFDA-ODA). The inclusion of fluorinated monomers in the polyimide structure is responsible to reduce their moisture absorption and dielectric constant. These polyimides increased susceptibility to chemical attack, making their use in multilayer fabrication. These formulations are commercially available in the polyamic acid form from Amoco under the trade name Ultradel. A slight modification in the polyimide chemical structure was done by using hexafluorodianhydride-aminophenoxy-biphenyl (HFDA-APBP). It is a well-known alternative formulation of an acetylene-terminated polyimide backbone, and it may be used when the polarization of interlayer is an important parameter. The cross-linking between low molecular weight oligomers during the curing through the acetylene end groups resulting in a high temperature, high molecular weight dielectric. It is suggested during the curing, cross-linking of the acetylene groups occurred without evolution of water and achieved excellent planarization by using low molecular-weight oligomers and high solubility resultant oligomers in solvents such as n-methyl pyrolidone (NMP). The polyimide poly(azomethine)s [6], polyimides [7], poly(amideimide)s [8] are soluble in organic solvents (chloroform, N,N-dimethylacetamide) because they contain pendent pentadecyl chain and these imides are synthesized using 4-pentadecylbenzene-1,3-diamine (PBDA). It is reported that in pyridine and lithium chloride medium, poly(ester-imide)s were derived from 2,2-bis(4-trimellitimidophenoxy)biphenyl,2,2-bis(4-trimellitimidophenoxy)-1,1-binaphthyl and aromatic dihydroxy compounds [9].

The nano-foam polyimides are synthesized, which have pores of size in nanometres [10] and prepared from thermally labile poly(propylene oxide) (PO) with thermally stable PMDA. The degradation products are diffused from the films when the produced foam is affected by thermolysis that pull out the pore size and shape from the initial polymer morphology. The dielectric constant of the polymer can be reduced by introducing air into the pores. The challenge of polymer scientists is to introduce the polyimides into thin film in a multilayer process. The polyimides are characterized using different sophisticated instruments such as infrared (IR) spectroscopy, nuclear magnetic resonance (NMR) spectroscopy, X-ray diffraction and thermogravimetric analysis (TGA).

2.1.2. *Polynitriles*

The nitrile polymer is composed of three monomers, namely acrylonitrile, butadiene and carboxylic acid. The properties of the polymer depend on the composition and attachment of monomers. The monomers are contributing individual ability to control the overall properties of the polymers. The presence of large occupying groups in the structure will increase the adhesion behaviour, stability and low dielectric constant. The presence of the monomer acrylonitrile and its polar nature will show reluctance to diffuse in many solvents, chemicals such as fats, hydrocarbon oils. The butadiene monomer improved flexibility, softness of the polymer and it is responsible for increasing elastic quality during vulcanization process with sulphur and accelerators. Polynitrile aromatics are prepared via palladium-catalyzed cyanation of aryl bromides with potassium hexacyanoferrate (II) [11] and plasma-polymerized 4-biphenylcarbonitrile (PPBPCN) thin films [12].

2.2. Silicon containing dielectric material

Silicones containing polymers are made up of the siloxane unit means alternating chain of silicon and oxygen atoms, combined with carbon and hydrogen. The materials have excellent heat resistant properties, and they are used as insulators, lubricants, adhesives and sealants in electrical, thermal, and pharmaceutical industries. The polymer properties in terms of resistance to oxidation can be modified by incorporating silicone compounds such as clays, silicates, silica, polyhydrogen silsesquioxanes (PHSSQ), and polymethylsilsesquioxane (PMSSQ) covalently bonded with the polymer matrix [13–15]. It was found out that solid polymer dielectrics have low thermal conductivity. So it is necessary to improve their conductivity without disturbing other electrical properties. So the studies suggested, introducing fillers that work as an insulator (silicon dioxide, SiO_2) increased the compatibility as well as adhesion behaviour of the polymer. Under electric stress, the polymers that contain silicon oxide or silicon carbide are more resistant to partial discharge, because the erosion depth reduced as compared to polymers without fillers [16–19].

2.3. Epoxy resins

Flexible thermosetting resins are known as polyiodides and are formed by copolymerization of an epoxide with another compound having two hydroxyl groups. Epoxy resin was first discovered by Pierre Castan, a Swiss chemist in the year 1938. The epoxy resin can be used in different fields such as electrical, electronics, paints, coating, engineering and aerospace because it has excellent properties of chemical resistance, mechanical strength, electrical insulation as well as excellent bonding with various curing substances [20, 21].

Epoxy resins can be used to clean the switches and prevent shorts in a various circuits, transistors and printed circuit boards with the help of their insulation properties. The epoxy resins are used in sophisticated electronics to increase the thermal resistance, capability to self-extinguish in case of emergency and physical flexibility. Epoxy-resin systems are thermosets, so it may be used in rotating machines, ground wall of generator, insulators and transformers [17, 22–24]. At elevated temperatures, epoxy resins are cross-linked through the polymerization reaction. The higher glass transition temperatures and strength of the resin system after curing increased more at higher temperature than room temperature [25]. The epoxy-silica composite is most popular resin due to its different properties such as enhanced partial discharge resistance [26, 27], higher volume resistivity [28], lowered thermal expansion [27], improved electrical breakdown property [27, 29], reduced erosion depth [27, 30], enhanced mechanical properties [27] and improved thermal conductivity [31]. The long-term break down degradation process of epoxy resin is improved by increasing treeing time to breakdown and incorporating small amount of zinc oxide or aluminium oxide nanoparticles [32–34]. Under constant AC voltage (10 kV to 1 kHz), the resin breakdown time will be double after addition of small amount of organically modified layered silicates than the base resin at 20°C and it will be six times at 80°C. At 145°C, the composite had breakdown time more than 20,000 min and the base epoxy resin had breakdown time of 280 min [18].

The epoxy resins are cross-linked with carboxyl or anhydride groups to make excellent resistance property, and the product is very useful as a coating in industry. The outstanding physical and chemical properties can be achieved by combination of glycidyl ether functional bisphenol A resins and carboxyl terminated polyester [35] but glyceryl ester functional acrylic polymers used 1, 12 dodecanedioic acid as a cross linker for powder-coating system [36]. The epoxy group reacted with the carboxyl group without a catalyst but the reaction rate can increase in the presence of a catalyst.

3. Polymers used in high-voltage applications

Basically, polymeric materials can be found naturally or synthetically thanks to chemical processes conducted in laboratories. Paper, wood, wax, leather, cellulose and rubber are examples of natural polymers; while the phenol formaldehyde commonly known as Bakelite is a good example of a polymer synthesized in laboratory. In several applications, simple polymers with some additives can be used for power cables, transformers, insulators and rotating machines [37, 38]. In general, the polymer used in high-voltage applications can be divided into three categories, according to their properties, such as thermal characteristics, physical and chemical structures, mechanical and electrical behaviours, among others (see **Table 3**) [37–40]:

1. **Thermoplastic polymers** are defined as plastics that soften and become supple on heating and solidify back on cooling [27]. The heating and cooling cycle within certain temperature limits can be applied to these materials several times without affecting their properties and without the need of any earlier chemical processing [41]. The synthetic thermoplastic polymers mostly used in insulation systems of electric machines and equipment are: polyethylene (PE), polyvinylchloride (PVC), polypropylene (PP) and polyamide (PA). Likewise, depending on their transition temperature characteristics, this type of polymers can be divided into two large classes: amorphous and crystalline [42].

   Polyethylenes are the most common commodity polymers that have found applications in household items, packaging, electric insulations, medical applications, automotive applications, containers, etc. Polymerization of the ethylene monomer gives linear or branched polyethylene that is mainly thermoplastic in nature depending on the manufacturing conditions. These conventional thermoplastic polyethylene plastics can be subjected to cross-linking to become a thermoset polyethylene called cross-linked polyethylene (XLPE or XPE) with different or improved properties. Thus, XLPE is linear or branched polyethylene that has long chains inter-connected to form three-dimensional network structures that improved density, chemical resistance and reduced the flow of the polymer. The polyethylene (PE) is cross-linked by chemical processes (peroxide, silane) and radiation energy (electron beam). The main advantage of the radiation process to chemical processes is that there is no requirement of any other chemicals to be used with the original polymer compound.

2. **Thermoset polymers** are materials cured by heat or chemical reaction and become infusible and insoluble. This process, leading to the formation of a network structure, is also

Type	Material	Application
Thermoplastics	Polyethylene (PE)	Cables
	Polyvinylchloride (PVC), polypropylene (PP), polyamide (PA)	Transformers
Thermosets	Epoxy impregnant:	Rotating machines
	Polyester-resin	
	Phenol-resin	
	Silicon-resin	
	Epoxy resins	
Elastomers	Silicone	Polymer insulators
	Ethylene propylene rubber (EPR)	Shed materials
	Ethylene propylene diene monomer (EPDM)	

Table 3. Types of polymeric insulation.

known as setting of the polymer [43]. The main characteristic of these materials is that they have a permanent irreversible polymerization, this occurs, because they have a cross-linked molecular structure and are formed in two polymerization stages. The first stage is formation of a polymer with linear chains. The second stage is the final cross-linked structure [41–43]. The thermal, mechanical and electrical properties of the thermoset polymers may vary considerably depending on their application. After the cross-linking stage, the material does not soften on reheating; instead it becomes thermally more stable compared to thermoplastic materials. From a practical point of view, polyester-resin, phenol-resin, silicon-resin and epoxy resins (widely used in rotating machines) can be considered as examples of cross-linked polymer resins [27].

3. **Elastomer polymers** are flexible polymers that comprise a low cross-link density and generally have a low Young's modulus and high-failure strain compared to other materials. Among the main elastomers, polymers commonly used in electrical applications are silicone, ethylene propylene rubber (EPR) and ethylene propylene diene monomer (EPDM). The elastomers such as silicone, EPR and EPDM are major elemental materials for polymer insulators [37–40].

Thermoplastic, thermoset and elastomer polymers can be used for several types of ageing mechanisms when they are used as insulation systems of electrical assets and subjected to electrical field.

4. Ageing in polymer insulation systems

The ageing process present in the insulation systems of an electric machine or equipment can be defined as the irreversible changes of their intrinsic properties due to action by one or several degradation mechanisms or factors [44]. These mechanisms are imposed by the operation, environment or test that influences the performance of the insulation material. In this sense, the polymer materials used in the insulation systems can be subjected to various kinds of stresses, even, under normal operating conditions. According to the literature, these mechanisms or degradation factors can be classified as thermal, electrical, ambient and mechanical (see **Figure 1**) [45–47].

Over time, all these ageing mechanisms tend to degrade the material and taking it to the final loss of its insulating properties and therefore the total breakdown of the material. The ageing caused by only one mechanism is denoted as single factor. When there are several mechanisms, either simultaneously or sequentially applied, it is called as a multi-factor stress. Below, each of these degradation mechanisms that may influence the lifetime of a polymeric material is described [48–50].

1. *Electrical stress*: The presence of non-uniform fields in polymeric insulation is a necessary condition for dielectric ageing under normal voltage and over-voltage conditions. When high-voltage is applied space charge is formed inside of the dielectric material. The space charge distorts the electric field distribution in the insulation and even in inter-facial regions for the case of composite insulation systems. This phenomenon occurs due to the

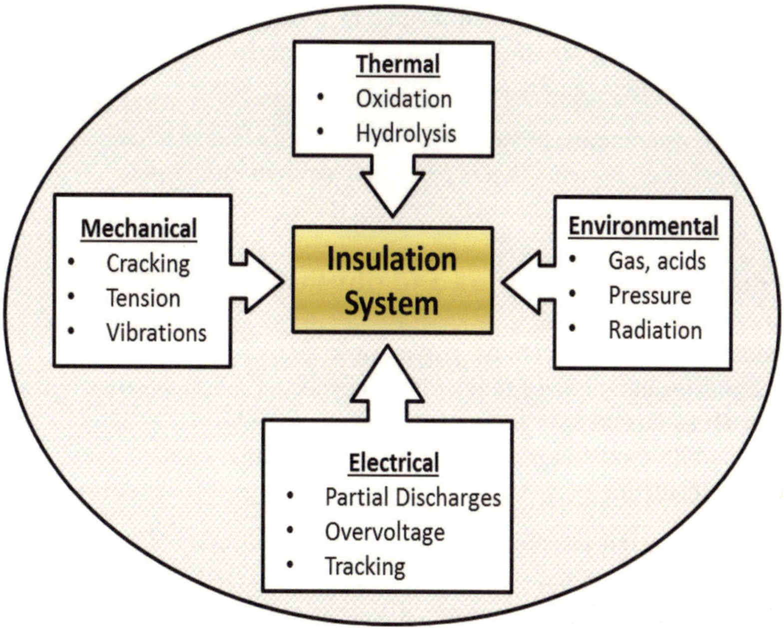

Figure 1. Degradation mechanisms or factors.

presence of air and other gases in the form of vacuoles within the polymeric material. All these factors facilitate the presence of a series of chemical changes thanks to an electronic and ionic bombardment of high energy that breaks chemical bonds and erodes isolation when subjected to continuous electrical stresses [38, 40].

2. *Thermal stress*: Can be considered another important stress factor in polymer insulation systems, since the increased temperature and heat cycles accelerate most of their degradation processes such as oxidation, hydrolysis, chemical attack and mechanical creep. Hot spots may arise due to ohmic losses in conductors, these tend to dissipate through the surrounding insulation. Finally, as a result of the thermal stress, the material can be degraded until it significantly lose its mechanical properties [38, 47].

3. *Environmental stress*: Can be identified as one the main reasons of failure in polymeric insulations. Among the main factors of environmental stress, the authors would like to highlight: sunlight, gamma ray and X-ray radiation, salt and dust contamination, acid rainfall, and even pressure. Thanks to these factors, many polymers commonly used as insulation on different equipment and electric machines begin to accelerate their ageing processes due to the occurrence of phenomena such as tracking, treeing, cracking, decomposition, pitting, dry-band arcing, erosion or oxidation, among others [48].

4. *Mechanical stress*: For most of the electrical equipment, mainly electric machines, the mechanical stress can appear in the form of compression, tension, bending, vibration and impact. All these factors act directly on the insulation, leading in many cases to fissures in the bulk of the material or at interfaces between materials of the insulation system.

Consequently, the insulation system begins to weaken product of these new spaces of lower dielectric strength, leading the insulation to failure. Likewise, other mechanisms of ageing can start acting simultaneously [49].

All above types of stresses can activate a typical and dangerous mechanism that can degrade the polymeric insulation material that is partial breakdown discharges.

5. Partial discharges: description and physical fundamentals

As mentioned in the previous sections, polymeric insulation systems associated with electrical machines and insulated wires may suffer unexpected failures associated with continuous stresses such as mechanical, thermal, electrical and environmental ones [44–46]. These stresses, after a while, tend to age and degrade the insulation system, leading to the permanent loss of their insulating properties and therefore to a possible breakdown.

In the pre-failure stage, it is possible to detect ionization processes of low energy, i.e. in small vacuoles trapped within the material or on the surface of a contaminated dielectric, wherein highly divergent electric fields are present. These processes are so-called partial discharges (PD) and can be considered as an important indicator of the overall condition of the insulation system [50, 51].

A partial discharge is a phenomenon of dielectric breakdown (low energy), limited to a region of the insulating medium between two conductors that are under different potentials [52]. The discharge location is due to an increase of electric field in a region that is relatively small compared to the overall dimensions of the insulating medium, or by the presence of a medium of lower dielectric strength in the insulation system. In some cases, both circumstances may occur simultaneously. This region must be completely or partially in the gaseous phase and may correspond, for example, to solid insulation occlusions, bubbles formed by vaporization of a liquid or gases surrounding conductive tips with small bending radius. Therefore, gases such as air, H_2, O_2, CO_2, etc. can be considered as enabling environments where PD may be originated [52].

According to the above, it can be considered that the partial discharge phenomenon is based on disruption processes of a gaseous dielectric, whose theories are much consolidated [53, 54]. From its study, it can be concluded that there is PD activity in an insulation system if there are two conditions happening simultaneously:

1. Presence of a sufficiently high-electric field: necessary to ensure an electrical breakdown in the area with lower dielectric strength or volume portion containing the gas. This condition also strongly depends on the pressure at which the gas is subjected, defined by the Panchen's Law [53, 54].

2. Free electrons available: necessary to ionize other neutral molecules with its acceleration in order to create new electron-ion positive pairs resulting in an avalanche process within the electric field.

In **Figure 2**, the applied voltage (U_a) to the whole insulation system and applied voltage in the area of greatest divergence of the electric field (U_c) and lower dielectric strength are represented. When the Uc voltage reaches the ignition voltage Ui^+, also called as partial discharge inception voltage (PDIV) in the electrically weaker volume, a discharge occurs, and immediately, U_c voltage drops sharply to a value (U_e) called extinction voltage discharge (DEV), thereby forming a transient current. After discharge, the gas in the cavity is subjected to progressively increasing voltage that increases trying to follow it. The increased voltage causes another discharge when it again reaches the level of the ignition voltage Ui^+ [55].

This process is repeated continuously in the positive and negative half cycle, generating through the insulation a current pulse whose behaviour is classified as stochastic. This character of partial discharges lies in the nature of initiation and evolution of the discharge. As it is seen above, both processes are strongly determined by quantum or mechanical-statistical aspects (appearance of free electrons, kinetic energy of electrons in collisions, etc.), so that, PD measurement should always be considered the probabilistic nature of the phenomenon [56].

According to this behaviour, PD activity will produce displacement of short duration currents (0.1–10 ns) in the insulation. These current pulses can be measured by the suitable detection circuits and generally are acquired and displayed superimposed on the applied voltage in so-called phase-resolved partial discharge (PRPD) patterns [55, 57].

As it is mentioned above, PD measurement allows detection of defects in insulation systems generated by multiple ageing agents. Moreover, apart from being a result of other mechanisms of degradation, PD activity will be an additional ageing process itself, since they can

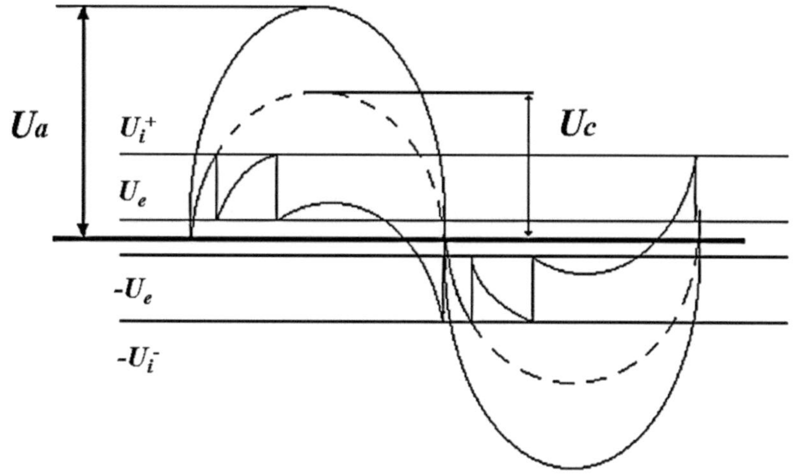

Figure 2. Partial discharges in a cavity subjected to alternating voltage.

continuously provoke [58]: (1) chemical attack generated by chemical reactions between oxygen, ozone and free radicals on the surface of the imperfection; (2) physical attack on the material due to electronic and ionic high-energy bombardment, which breaks chemical bonds and erodes the insulation system; (3) deterioration of imperfection surface caused by ultra violet (UV) radiation and (4) additional heating of the dielectric and power losses.

The combination of these ageing effects can accelerate the breakdown of the insulation system of any electrical asset (high-voltage cables, power transformers, rotating machines, etc.). Therefore, PD activity detection, tracking, identification of the source type, and its evolution over time are of great importance to establish the insulation condition of electrical equipment and to be able to apply an optimum maintenance.

5.1. Types of partial discharge

Basically, the PD phenomenon can be divided into three basic categories according to their origin: internal, surface and corona [59]. However, the vast experience developed in PD measurement in many types of high-voltage equipment allows detecting, with higher accuracy, other concrete defects. The PD type identification is important because not all PD sources are necessarily harmful to the electrical equipment, as will be seen later.

5.1.1. Internal PD

Internal PD may occur in vacuoles with low dielectric strength, encrusted foreign particles, metal bumps and internal cracks in the material or in the nearby border to one of the electrodes (see **Figure 4a**). The origin of such PD depends on the defect geometry, the ageing of the material due to the different stresses to which it may be subjected (mechanical, thermal and electrical), the electric stress strength in the cavities of the material due to difference between permittivity of the two media (insulation and imperfection), and free electrons from previously destroyed chemical bonds or by ionizing radiation [59].

The number of discharges per cycle is generally affected by the appearance of space charge in the bulk material, which explains that their appearance is highly probable for the phase in which the slope of the applied voltage is maximum (see **Figure 3**). In **Figure 4a**, it is shown the physical process of internal PD. It is observed that when the DIV is exceeded, it is probably that an ionization occurs in imperfection (depending on the presence of a free electron), causing the appearance of new charge carriers (electrons and ions) resulting from the breakdown of bonds of neutral gas molecules. Therefore, after a PD starts, the charges appearing distributed within the cavity depending on their polarity generating clusters of space charge on its walls. This space charge causes an induced electric field inside the insulation (E_q), which overlaps the electric field generated by the applied voltage (E_i). Initially, the induced field opposes the applied field, reducing the total applied field. However, in polymers subjected to alternating voltages, the applied electric field changes its direction in each cycle, which may cause the total field ($E_t = E_i + E_q$) at the instant of polarity change saw intensified, as the field induced by the accumulated charge can maintain its polarity until it returns to redistribute after another PD. This effect of the induced field can be mitigated by eliminating the space

http://dx.doi.org/10.5772/67091

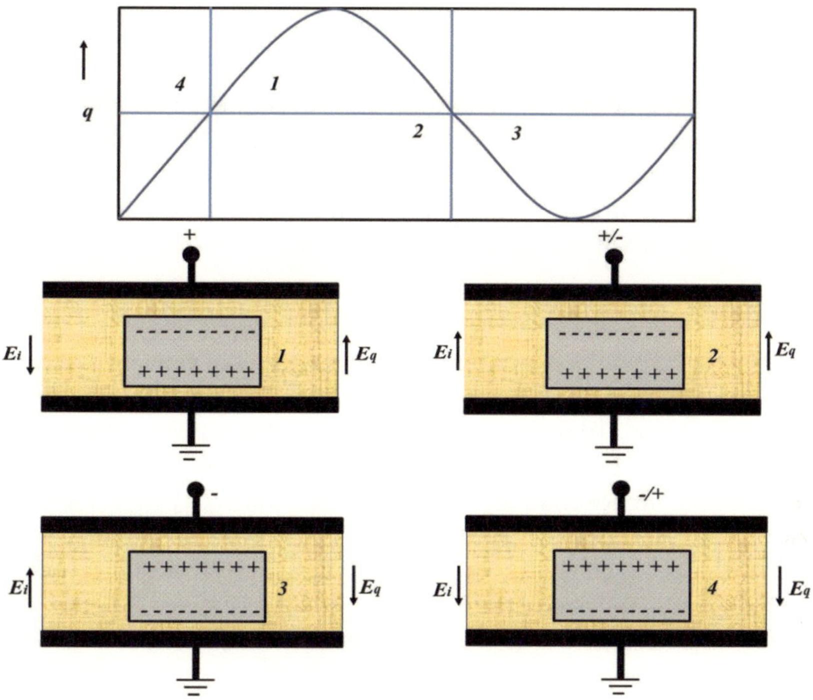

Figure 3. PD produced in a cavity of a dielectric result of the superposition of the electric field induced by charge E_q and electric field applied by the alternating voltage E_i.

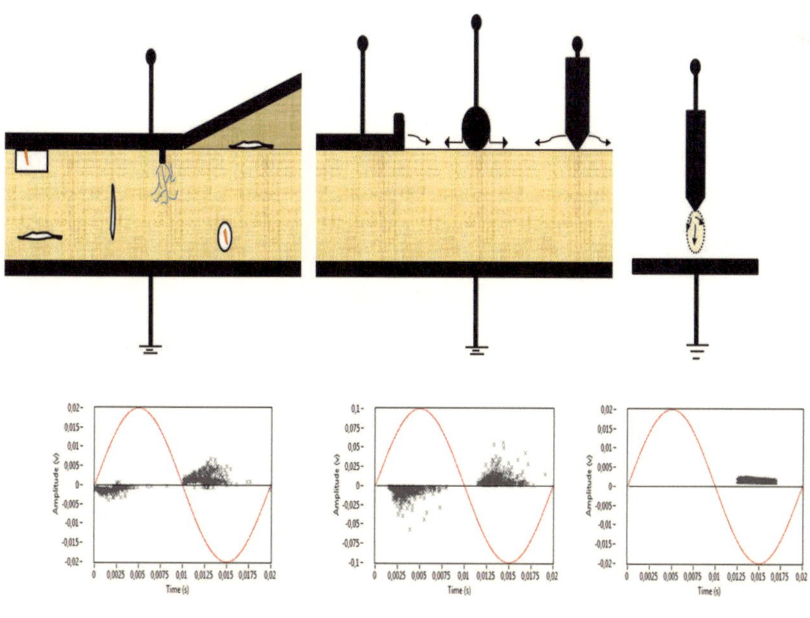

Figure 4. PD example with corresponding PRPD pattern (a) internal PD, (b) surface PD and (c) corona PD.

charge of the vacuole walls due to increased conductivity of the same or by a repetition rate of discharge too low [60, 61].

In general, the presence of internal PD sources can be considered as critical to the insulation system, because its activity is concentrated in smaller and weak points of the material. Besides, in the case of having a persistent activity they can erode the material surrounding the vacuole to completely destroy it. For these reasons, it is necessary to well identify their main electrical characteristics as the voltage ignition in polymer's cavities.

5.1.2. Surface PD

Surface PD occurs at the interfaces between dielectric materials or between the conductive element and the insulation when any type of contamination exists on the material surface (rain, snow, moisture, pollution, salinity, etc.). This is because the presence of tangential electric field components of large intensity can originate discharges that are extended beyond their origin point along the surface of the insulation in order to follow a path to the electrode with a lower potential. In practice these types of discharge are presented in cable terminals, coil heads rotary machines, contaminated insulators and bushings, among other sources. Various examples of configurations that give rise to this type of shock observed in **Figure 4b**. Although this type of discharges can be harmful, their detection could help to locate the source and mitigate its presence, for example, cleaning the contaminated surface or by applying a layer of silicone (hydrophobic material) insulator.

These types of partial discharges occur in phases with higher applied voltage and present, as internal PD, great variability in the detected magnitude (are events conditioned by the state of the insulating surfaces, which can changing), being more likely, discharges with lower magnitude, see **Figure 4b**.

5.1.3. Corona PD

As it is shown in **Figure 4c**, corona PD occur around a sharp conductive material or with a small radius of curvature subjected to a strong electric field within dielectric liquid or gas [59]. The inhomogeneity caused by high-electric field gradients make that in the vicinity of the tip will produce the ionization of the surrounding gas, and therefore, a local insulation breakdown in the area of high-electric field strength.

In volume material farthest from the sharp electrode, the electric field is reduced and ionization is extinguished. This phenomenon is typical in transmission lines, provoking in them energy losses, radiofrequency interferences, even audible, and degradation of polymeric insulator by the appearance of nitric acid [62]. In some cases, these discharges can be caused in loose threads in high-voltage connections, which are not an immediate threat for the electrical assets. Finally, if corona PD sources cause is detected, it can be easily removed.

In order to determine the electric-field effect on polymer insulation systems, next section shown experimental results for PD measurements in insulated underground cables.

6. Partial discharges in polymeric insulation: underground cables

Cross-linked polyethylene (XLPE) is commonly applied as the insulation material for high-voltage underground cables. XLPE possesses excellent dielectric properties and can withstand high-voltages up to 380 kV. As stated previously, one major PD defect in the HV insulation is the void. It is necessary to understand possible internal PD activity over time in order to ascertain whether different void PD patterns vary in uniqueness and characteristics. It is also important to understand different void degradation characteristics and how they affect the lifeline of the HV apparatus. Discharges in voids are known to have unpredictable behaviour [4]. Discharges and PD amplitude in a void can sometimes increase or decrease and in some cases even disappear. Discharges impacting at the void surface depositing conducting layers (e.g. oxalic acid and water) around the void boundary due to chemical decomposition reactions of the polymer. These conducting charge carriers get trapped in the surface and may change the PD mechanism due to variations in the boundary conditions. In a void, streamer discharges prevail over the first hour of stressing and afterwards these normally change to Townsend discharges as a result of the change in resistivity of the surface due to sustained charge carriers [5].

A number of research investigations have analysed PD activity within voids in polymeric insulation systems [5, 63, 64], but few have analysed voids in polyethylene-terephthalate (PET) insulations [4, 65]. PET is a thermoplastic polymer resin belonging to the polyester family. It has mechanical strength and an excellent moisture- and water-protective system. In this section, experimental results void developments in PET will be reported. Specifically, experimental results involving dielectric bounded an electrode-bounded cavity in PET over a 7-h stressing stage will be reported. This is important in order to understand the deterioration mechanism of PET over hours of degradation. In conducting the experiments and to ensure that the experimental set-up is discharge-free prior to the samples with voids being investigated, PET layers without any voids were primarily stressed to confirm discharge free environments for PET layers.

6.1. Dielectric bounded cavity

For the dielectric-bounded void, experiments were conducted using a set of samples containing dielectric-bounded void in PET. Up to nine PET layers are created for each experiment and the void is located within the centre of the PET layers (see **Figure 5**). The inception voltage was approximately 2.82 kV and all measurements were taken at approximately 3.6 kV. PD data was capt ured over 50 and 250 power cycles from the start up to 7 h continuous stressing.

Figure 6 shows typical ϕ-q-n (phase-amplitude-number) patterns for the dielectric-bounded void at 1 h and 7 h periods captured over 50 and 250 cycles, respectively. It can be observed that discharges are centred between 20° and 110° in the positive half cycle and between 150° and 270° in the negative half cycle. Also, it appears there is no significant maximum amplitude change between the 1 h and 7 h captures for both the 50 and 250 cycle captures in

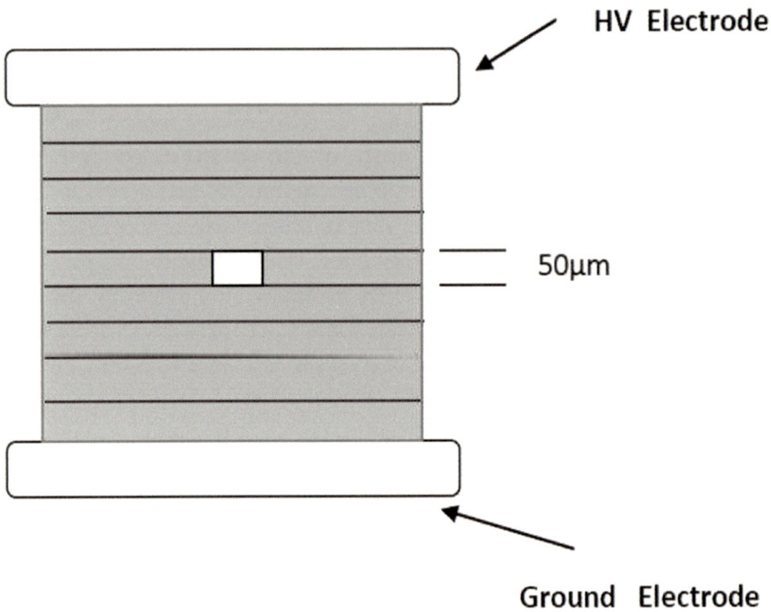

Figure 5. Sample set-up showing a dielectric bounded void at the middle of the PET layers.

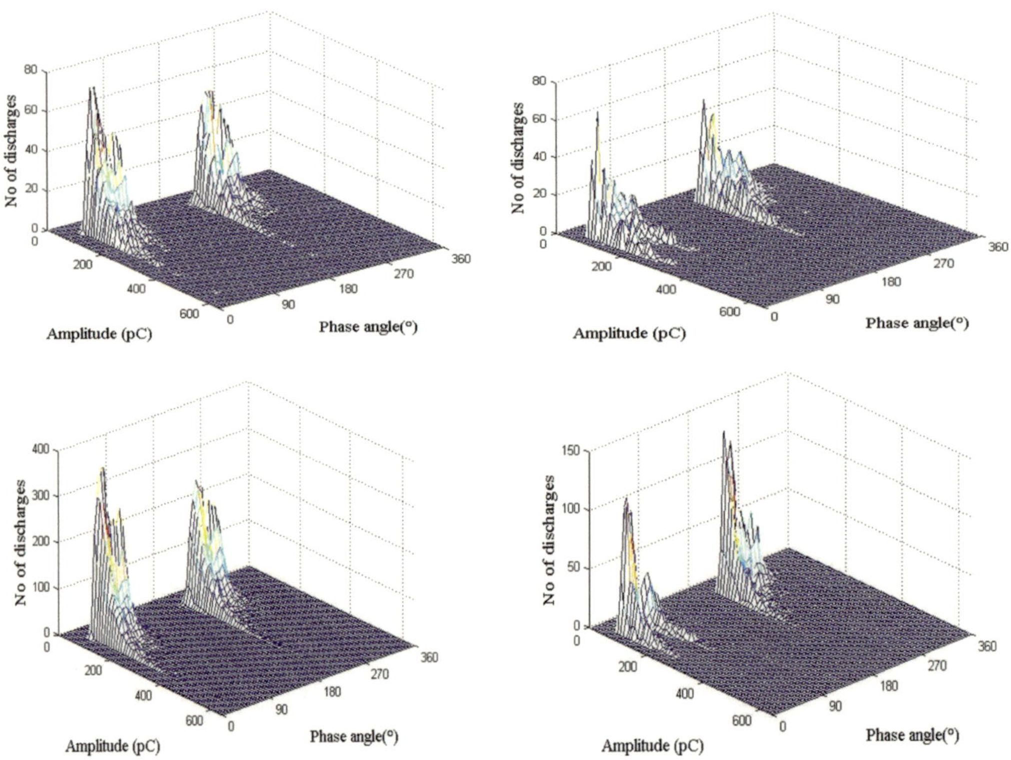

Figure 6. Series of φ-q-n plots for dielectric bounded void captured at (a) 1 h, 50 and 250 cycles, (b) 7 h, 50 and 250 cycles.

the dielectric-bounded void (see example **Figure 7a** and **b**). This indicates that the void is still at the initial degradation period and probably the insulated cable has enough lifetimes to be in service.

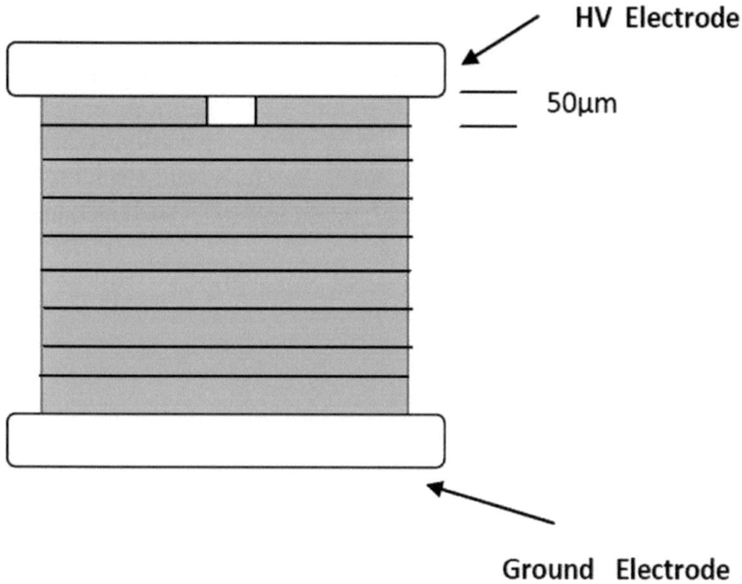

Figure 7. Electrode bounded void.

6.2. Electrode bounded void

Discharges from an electrode-bounded void are studied by placing a 10 mm diameter void layer to the top-most PET layer adjacent to the HV electrode (see **Figure 7**). This is to ensure that the void has an electrode surface at one side and PET surface on the opposite side. Similar to the dielectric-bounded void experiments, nine PET layers are used, including the one possessing the cavity. For all the experiments, a voltage of 3.6 kV is applied and ϕ-q-n patterns captured at 15-min intervals over a period of 7 h.

Examples of the φ-q-n patterns are shown in **Figure 8**. In comparison to dielectric-bounded voids considered in the previous section, asymmetry between the discharges in the positive and negative half cycles is visible in the electrode-bounded cavity consistent with the general literature [66]. Gulski's conclusions are based on a 20-min test but in this work discharge patterns are captured up to what appears to be initial stages of deterioration of the void (i.e. over a 7-h period). The asymmetry between the two half-power cycles can be explained as follows. Gulski [66] suggested that during the positive half-power cycle, the residual charge on the dielectric surface produces a high number of instigating electrons, while in the negative half-power cycle, the initiating electron may be released from the electrode surface itself or from background radiation.

7. Conclusion

This chapter presents a comprehensive study on the general structure of polymers, their properties and applications. Particular interest is on the application of polymers as insulating materials for high-voltage apparatus such as transformers, electrical machines and underground cables. The effect of electric field distribution on polymeric insulations subjected to partial discharges have been reported and analysed. Epoxy resin and XLPE polymers are the widely applied for high-voltage applications though it is necessary to estimate and understand their lifetime. Internal partial discharges in polymer voids can create a discontinuity of electric field in the insulation system and a continued activity can lead to total breakdown of the high-voltage insulation. Therefore, it is important to understand and obtain new information regarding novel polymeric materials such as nano-polymers that can possibly outperform currently used ones.

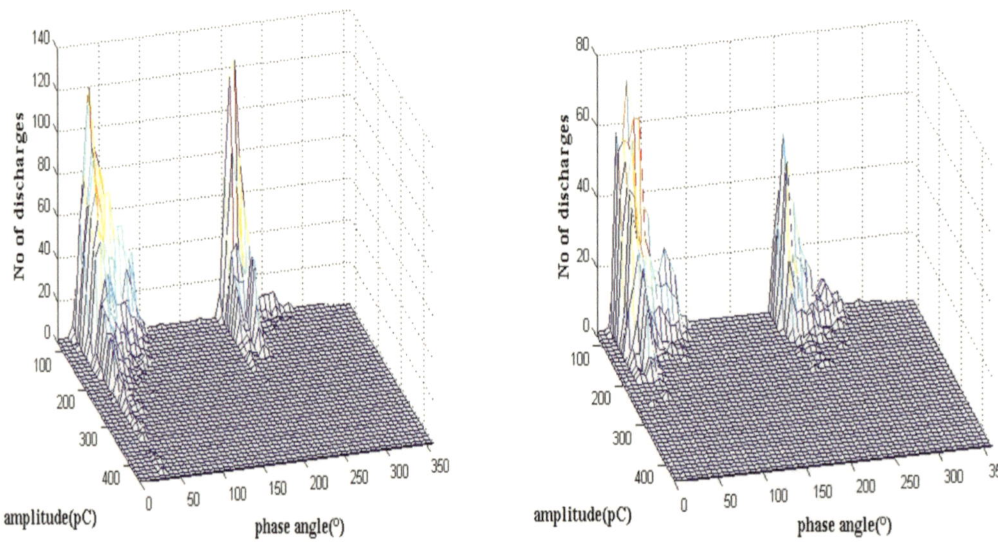

Figure 8. Series of φ-q-n plots for electrode bounded void captured at (a) 1 h and (b) 7 h.

For practitioners in the field, it is necessary to design correct information system for electrical equipment, i.e. by using the appropriate polymer for solid insulation, avoiding metallic sharp components and any other conducting path in order to prevent any type of external PD, such as surface and corona. Breakdowns of polymers and other kinds of insulation systems can be warned by measuring and identifying PD activity, mainly internal PD sources by using condition-monitoring techniques.

Acknowledgement

The authors acknowledge the support of the Chilean Research Council (CONICYT), under the project Fondecyt 11160115.

Author details

Sk Manirul Haque[1], Jorge Alfredo Ardila Rey[2], Abdullahi Abubakar Masúd[3]*, Yunusa Umar[1] and Ricardo Albarracin[4]

*Address all correspondence to: abdullahi.masud@gmail.com

1 Department of Chemical and Process Engineering Technology, Jubail Industrial College, Kingdom of Saudi Arabia

2 Department of Electrical Engineering, Technical University Federico Santa María, Santiago de Chile, Chile

3 Department of Electrical and Electronic Engineering Technology, Jubail Industrial College, Kingdom of Saudi Arabia

4 Universidad Politécnica de Madrid, Madrid, España

References

[1] Alfred S C and Charles E W Jr, Electric Capacitor and Method of Making the Same, US Patent 3252830 assigned to General Electric Company, NY May 1966.

[2] Milton H P, Electric Capacitor, US Patent 3311801 assigned to Intron International, NY, Mar 1967.

[3] Nalwa, H. Handbook of Low and High Dielectric Constant Materials and Their Applications; Academic Press: London, UK, 1999.

[4] Kochetov, R, et al. "Impact of postcuring and water absorption on the dielectric response of epoxy-based composites filled with MgO nanoparticles." 2011 Annual Report Conference on Electrical Insulation and Dielectric Phenomena (CEIDP). IEEE, 2011.

[5] Deligöz, H, et al. "A novel cross-linked polyimide film: synthesis and dielectric properties." Polymer 46.11 (2005): 3720–3729.

[6] Sadavarte, Nilakshi V, C V Avadhani, and P P Wadgaonkar. "Synthesis and characterization of new organosoluble aromatic polyamides and polyazomethines containing pendent pentadecyl chains." High Performance Polymers 23.7 (2011): 494–505.

[7] Sadavarte, N V, et al. "Synthesis and characterization of new polyimides containing pendent pentadecyl chains." European Polymer Journal 45.2 (2009): 582–589.

[8] Sadavarte, N V, et al. "Regularly alternating poly(amideimide) s containing pendent pentadecyl chains: synthesis and characterization." European Polymer Journal 46.6 (2010): 1307–1315.

[9] Behniafar, H, and S Haghighat. "Thermally stable and organosoluble binaphthylene-based poly(urea-ether-imide)s: one-pot preparation and characterization." Polymers for Advanced Technologies 19.8 (2008): 1040–1047.

[10] Carter, K R, et al. "Polyimide nanofoams for low dielectric applications." MRS Proceedings. Vol. 381. Cambridge University Press, 1995.

[11] Becker, M, A Schulz, and K Voss. "Synthesis of poly-nitrile aromatics via palladium-catalyzed cyanation of aryl bromides with potassium hexacyanoferrate (II)." Synthetic Communications 41.7 (2011): 1042–1051.

[12] Zhao, X-Y, et al. "Structural and dielectric properties of conjugated polynitrile thin films deposited by plasma polymerization." Thin Solid Films 516.23 (2008): 8272–8277.

[13] Aw, K C, et al. "Characterization of spin-on-glass very-low-k polymethylsiloxane with copper metallization." Thin Solid Films 504.1 (2006): 243–247.

[14] Singh, S K, et al. "Potential of Cat-CVD deposited a-SiC: H as diffusion barrier layer on low-k HSQ films for ULSI." Thin Solid Films 501.1 (2006): 318–321.

[15] Singh, S K, A A Kumbhar, and R O Dusane. "Maintaining Cu metal integrity on low-k IMDs with a nanometer thick a-SiC: H film obtained by HWCVD." Thin Solid Films 516.5 (2008): 785–788.

[16] Henk, P O, T W Kortsen, and T Kvarts. "Increasing the electrical discharge endurance of acid anhydride cured DGEBA epoxy resin by dispersion of nanoparticle silica." High Performance Polymers 11.3 (1999): 281–296.

[17] Henk, P O, et al. "Increasing the PD-endurance of epoxy and XLPE insulation by nanoparticles silica dispersion in the polymer." Nordic Insulation Symposium, Stockholm, Sweden. 2001.

[18] Imai, T, et al. "Evaluation of insulation properties of epoxy resin with nano-scale silica particles." Proceedings of 2005 International Symposium on Electrical Insulating Materials, 2005 (ISEIM 2005). Vol. 1. IEEE, 2005.

[19] Imai, T, et al. "Influence of temperature on mechanical and insulation properties of epoxy-layered silicate nanocomposite." IEEE Transactions on Dielectrics and Electrical Insulation 13.2 (2006): 445–452.

[20] Tanaka, T, Y Matsuo, and K Uchida. "Partial discharge endurance of epoxy/SiC nanocomposite." 2008 Annual Report Conference on Electrical Insulation and Dielectric Phenomena. IEEE, 2008.

[21] Singha, S, and M J Thomas. "Dielectric properties of epoxy nanocomposites." IEEE Transactions on Dielectrics and Electrical Insulation 15.1 (2008): 12-23.

[22] Zhang, C, R Mason, and G C Stevens. "Dielectric properties of alumina-polymer nanocomposites." CEIDP'05. 2005 Annual Report Conference on Electrical Insulation and Dielectric Phenomena, 2005. IEEE, 2005.

[23] Zou, C, J C Fothergill, and S W Rowe. "The effect of water absorption on the dielectric properties of epoxy nanocomposites." IEEE Transactions on Dielectrics and Electrical Insulation 15.1 (2008): 106–117.

[24] Luft, J H. "Improvements in epoxy resin embedding methods." The Journal of Biophysical and Biochemical Cytology 9.2 (1961): 409–414.

[25] Imai, T, et al. "Insulation properties of nano-and micro-filler mixture composite." CEIDP'05. 2005 Annual Report Conference on Electrical Insulation and Dielectric Phenomena, 2005. IEEE, 2005.

[26] Li, H, et al. "Synthesis and characterization of epoxy resin modified with nano-SiO 2 and γ-glycidoxypropyltrimethoxy silane." Surface and Coatings Technology 201.9 (2007): 5269–5272.

[27] Plesa, I, et al. "Properties of polymer composites used in high-voltage applications." Polymers 8.5 (2016): 173.

[28] Preetha, P, et al. "Electrical discharge resistant characteristics of epoxy nanocomposites." 2008 Annual Report Conference on Electrical Insulation and Dielectric Phenomena. IEEE, 2008.

[29] Kochetov, R, et al. "Thermal conductivity of nano-filled epoxy systems." IEEE Conference on Electrical Insulation and Dielectric Phenomena, Virginia Beach, VA, USA. 2009.

[30] Ding, H- Z, and B R Varlow. "Effect of nano-fillers on electrical treeing in epoxy resin subjected to AC voltage." 2004 Annual Report Conference on Electrical Insulation and Dielectric Phenomena, 2004. CEIDP'04. IEEE, 2004.

[31] Alapati, S, and M J Thomas. "Electrical tree growth in high voltage insulation containing inorganic nanofillers." Proceedings of the 16th International Symposium on High Voltage Engineering, Cape Town, South Africa. 2009.

[32] Danikas, M G, and T Tanaka. "Nanocomposites—a review of electrical treeing and breakdown." IEEE Electrical Insulation Magazine 25.4 (2009): 19–25.

[33] McLafferty, J J, and S L Wang. "1-methyl imidazole catalyzed carboxyl terminated polyester." U.S. Patent No. 4,910,287. 20 Mar. 1990.

[34] Victorius, C. "Thermosetting powder coating composition of an acrylic polymer, dodecanedioic acid, cellulose acetate butyrate and a cross-linking agent." U.S. Patent No. 4027066. 31 May 1977.

[35] Gitzen, W.H. (Ed.). "Alumina as a ceramic material." United States: American Ceramic Society. 1970.

[36] Steiner, CJ-P, D P H Hasselman, and R M Spriggs. "Kinetics of the gamma-to-alpha alumina phase transformation." Journal of the American Ceramic Society 54.8 (1971): 412–413.

[37] Raju, G G. Dielectrics in Electric Fields. Marcel Dekker, Inc., New York, 2003.

[38] Naidu, M S, and V Kamaraju. High Voltage Engineering. Tata McGraw-Hill Companies, Inc., New York, 1995.

[39] Tanaka, T. "Aging of polymeric and composite insulating materials. Aspects of interfacial performance in aging." IEEE Transactions on Dielectrics and Electrical Insulation 9.5 (2002): 704–716.

[40] Arora, R, and W Mosch. High Voltage and Electrical Insulation Engineering. Vol. 69. John Wiley & Sons, Inc., Hoboken, New Jersey, 2011.

[41] Xanthos, M, et al. "Thermoplastic composites from maleic anhydride modified post-consumer plastics." Polymer Composites 16.3 (1995): 204–214.

[42] Amin, S, and M Amin. "Thermoplastic elastomeric (TPE) materials and their use in outdoor electrical insulation." Reviews on Advanced Materials Science 29.1 (2011): 15–30.

[43] Agarwal, V K, et al. "The mysteries of multifactor ageing." IEEE Electrical Insulation Magazine 11.3 (1995): 37–43.

[44] Gill, P. Electrical Power Equipment Maintenance and Testing. CRC Press, 2008.

[45] Wang, L, et al. "Evolution of PD patterns in polyethylene insulation cavities under AC voltage." IEEE Transactions on Dielectrics and Electrical Insulation 19.2 (2012): 533–542.

[46] Stone, G C, et al. Electrical Insulation for Rotating Machines: Design, Evaluation, Aging, Testing, and Repair. Vol. 21. John Wiley & Sons, 2004.

[47] Gjerde, A C. "Multifactor ageing models-origin and similarities." IEEE Electrical Insulation Magazine 13.1 (1997): 6–13.

[48] Montanari, G C, and G Mazzanti. "Ageing of polymeric insulating materials and insulation system design." Polymer International 51.11 (2002): 1151–1158.

[49] Kimura, K. "Multistress aging of machine insulation systems." Annual Report, Conference on Electrical Insulation and Dielectric Phenomena, 1995. IEEE, 1995.

[50] Stone, G C, and V Warren. "Effect of manufacturer, winding age and insulation type on stator winding partial discharge levels." IEEE Electrical Insulation Magazine 20.5 (2004): 13–17.

[51] Okabe, S, et al. "Partial discharge-induced degradation characteristics of insulating structure constituting oil-immersed power transformers." IEEE Transactions on Dielectrics and Electrical Insulation 17.5 (2010): 1649–1656.

[52] Van Brunt, R J. "Physics and chemistry of partial discharge and corona. Recent advances and future challenges." IEEE Transactions on dielectrics and Electrical Insulation 1.5 (1994): 761–784.

[53] Kuffel, J, and P Kuffel. High voltage engineering fundamentals. Newnes, 2000.

[54] Khalifa, M. "High Voltage Engineering, Theory and Practice", Marcel and Dekker. Inc., New York, 1990

[55] Van Brunt, R J. "Stochastic properties of partial-discharge phenomena." IEEE Transactions on Electrical Insulation 26.5 (1991): 902–948.

[56] Krivda, A. Recognition of discharges: discrimination and classification. TU Delft, Delft University of Technology, 1995.

[57] Cigré, W G. "21-03," Recognition of discharges." Electra 11 (1969): 61–98.

[58] Morshuis, P H F. "Partial Discharge Mechanisms." Delft University Press 51 (1993): 1–4.

[59] Kreuger, F H. Partial Discharge Detection in High-voltage Equipment. Butterworth-Heinemann, 1989.

[60] Illias, H, G Chen, and P L Lewin. "Modeling of partial discharge activity in spherical cavities within a dielectric material." IEEE Electrical Insulation Magazine 27.1 (2011): 38–45.

[61] Cavallini A, Montanari G. C; "Effect of supply voltage frequency on testing of insulation system"; Dielectrics and Electrical Insulation, IEEE Transactions on, vol.13, pp. 111–121, Feb. 2006.

[62] A. L. Souza and I. J. S. Lopes, "Experimental investigation of corona onset in contaminated polymer surfaces," in IEEE Transactions on Dielectrics and Electrical Insulation, vol. 22, no. 2, pp. 1321–1331, April 2015.

[63] Morshuis, P H F, and F H Kreuger. "A relation between time-resolved discharge parameters and ageing." Sixth International Conference on Dielectric Materials, Measurements and Applications, 1992. IET, 1992.

[64] Álvarez, F. Garnacho, R. Albarracín, R. Granizo, M. Dong and J. Ortego, "Practical experience of insulation condition evaluation in an on-site HV installation applying a PD measuring procedure," 2016 International Conference on Condition Monitoring and Diagnosis (CMD), Xi'an, 2016, pp. 509–513.

[65] Morshuis, P H F. "Partial discharge mechanisms in voids related to dielectric degradation." IEE Proceedings-Science, Measurement and Technology 142.1 (1995): 62–68.

[66] Gulski, E. Computer-aided recognition of partial dicharges using statistical tools. TU Delft, Delft University of Technology, 1991.

Chapter 4

Investigating the Influence of Different Types of Nanoparticles on Thermal and Dielectric Properties of Insulation in Converter Transformer

Boxue Du

Additional information is available at the end of the chapter

http://dx.doi.org/10.5772/67432

Abstract

Converter transformer is the extraordinarily vital apparatus in the high-voltage direct current (HVDC). Transformer oil and oil-impregnated paper are considered as the essential parts of converter transformer, which has suffered various complex electric fields such as AC/DC composite electric field and polarity reversal. The occurrence of discharge or insulation failure in oil-paper insulation system will threaten the safety of the entire grid. Hence, efforts should be focused on investigating the charge behavior and improving the property of dielectric.

This chapter presents a study aimed at investigating the influence of different types of nanoparticles on thermal and dielectric properties along with the relation between them and the effects of switching overvoltage on the charge coupling behavior of oil-impregnated paper, and the surface charge behavior on double-layer oil-paper insulation under pulse voltage is investigated.

Keywords: converter transformer, oil-impregnated paper, nanoparticles, thermal conductivity, operation overvoltage, surface charge

INTECH
open science | open minds

© 2017 The Author(s). Licensee InTech. This chapter is distributed under the terms of the Creative Commons Attribution License (http://creativecommons.org/licenses/by/3.0), which permits unrestricted use, distribution, and reproduction in any medium, provided the original work is properly cited.

1. Introduction

In recent years, the utilization of an HVDC system is becoming an inevitable trend with the developments in the worldwide power industry and improvements in voltage levels [1–3]. Meanwhile, oil-paper insulation has been widely used in converter transformers as well as HVDC bushings, and its safety is closely related to the security of large power transformers even the entire grid. Usually, oil-paper insulation withstands DC voltage under normal conditions. When extreme states appear, it will be subjected to various overvoltages, and the overvoltage will cause the rise of the electric field of the oil-paper insulation in a short time. Operating impulse voltage is a kind of overvoltage that often occurs in converter transformers and HVDC bushings. For example, the switching of thyristors often appears in power electronic equipment, and overvoltage has a cumulative and repeated effect on oil-paper insulation damage. So, it is necessary to study the effects of operating overvoltage on oil-paper insulation. Actually, charge accumulation under overvoltage is one of the main reasons that cause aging and breakdown of oil-paper insulation. But the effect of operating overvoltage on the charge injection and dissipation is still unclear. Current researches mainly focus on effects of lightning pulse voltage on oil-paper insulation, but for operating overvoltage, the influence of its cumulative effect on oil-paper insulation needs further investigation. Furthermore, how DC and pulse voltages affect charge coupling of oil-paper insulation calls for more consideration and investigation.

Furthermore, the mineral oil applied in transformers fills up the pores in the fibrous insulation and performs dual functions of insulating materials and effective coolant. For thermal property, the oil carries away the heat generated during normal operation by conduction and convection [4]. Moreover, the oil can affect the dielectric property of oil-paper insulation system and thus influence the breakdown strength of the transformers. New type of transformer oil with enhanced dielectric property and thermal conductivity is in urgent need with the miniaturization and high voltage of electric devices processing. Most studies on nano-modified transformer oil were limited to the dielectric property or thermal property. Very few researchers considered investigating the relation between the thermal property and the dielectric breakdown strength. In this chapter, Boron nitride (BN) nanoparticles with high thermal conductivity and ferriferrous oxide (Fe_3O_4) nanoparticles as a typical magnetic oxide were dispersed into transformer oil to form different types of nanooil. Thermal properties including thermal conductivity and thermal diffusivity were measured with varying the mass fraction and temperature. Dielectric properties and breakdown strength under different temperature were measured to investigate the influence of temperature on the nanooil.

Therefore, this chapter will focus on the investigating the influence of different types of nanoparticles on thermal and dielectric properties along with the relation between them and the effects of different application conditions on surface charge behavior of oil-paper insulation to understand and analyze the transportation of charge in the oil-paper insulation and the effects of voltage amplitude, pulse frequency, numbers as well as polarity on the charge accumulation and dissipation.

http://dx.doi.org/10.5772/67432

2. Thermal conductivity and dielectric characteristics of transformer oil filled with BN and Fe_3O_4 nanoparticles

2.1. Experiments

In this research, the average grain diameter and surface to volume of the BN nanoparticles are 50 nm and 43.6 m^2/g, while they are 20 nm and 66 m^2/g for the Fe_3O_4 nanoparticles. The nanoparticles were dried to decrease the adsorbed water content before dispersion. The transformer oil used was purified through a vacuum oil filter. The two types of nanoparticles with the mass fraction of 0.05, 0.1 wt% were dispersed into the filtered oil by a stirrer and ultrasonic vibrated to promote dispersion and reduce possible agglomeration. All the samples were degassed before measurements. The nanoparticles were dispersed uniformly in the base oil.

2.2. Effect on thermal conductivity

The THW method is employed to measure the thermal parameters. Thermal conductivity of the two types of nanooil and the pure oil is shown in **Figure 1**. The nanooil with BN nanoparticles shows more obvious enhancement compared to the oil dispersed with Fe_3O_4 nanoparticles under the same mass fraction and temperature. The BN-modified nanooil at the mass fraction of 0.05 wt% reaches the thermal conductivity of 0.1184 W m^{-1} K^{-1} when the applied

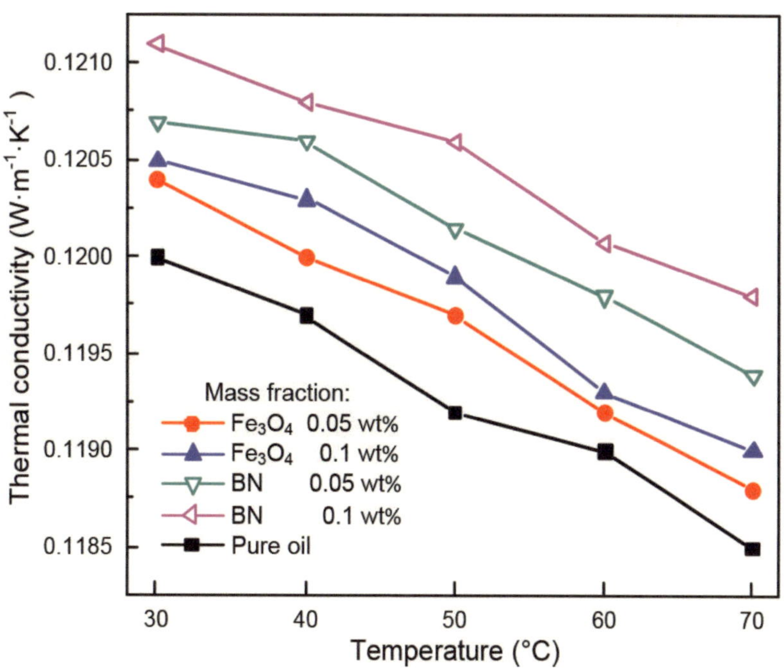

Figure 1. Relation between the thermal conductivity and the temperature with the different mass fractions.

temperature is 70°C which is even higher than the oil modified with Fe_3O_4 nanoparticles at the mass fraction of 0.1%. Thermal conductivity increases with increasing the mass fraction and decreases with increasing temperature for each type of the nanooil. The thermal conductivity shows linear decline tendency in the variation of temperature as measured.

It is considered that the interfacial region due to the addition of nanoparticles and the ballistic phonon transport among nanoparticles contribute to the considerable improvement. The liquid molecules close to the nanoparticles form a nanolayer which functions as a bridge linking the additional nanoparticles and the liquid molecules, thus causing the increase of thermal conductivity [5]. Moreover, Keblinski presented a ballistic phonon transport theory that the ballistic phonons originated in one particle can spread to an adjacent particle and contribute to the remarkable enhancement much compared to Brownian motion even at extremely low mass fraction [6]. The higher thermal conductivity of the BN-modified oil compared with the Fe_3O_4-modified oil is expected to attribute to the higher thermal conductivity of BN nanoparticles.

Figure 2 shows the temperature-dependent thermal diffusivity of the pure oil and the nanooil modified with BN and Fe_3O_4 nanoparticles, respectively. With the mass fraction of 0.1%, the increase in thermal diffusivity is observed to be 8.7% for the BN-modified oil and 2.6% for the Fe_3O_4-modified oil at the temperature of 30°C, respectively. As shown in **Figure 2**, the thermal diffusivity of the oil performs nearly the same decline slope with varying the temperature

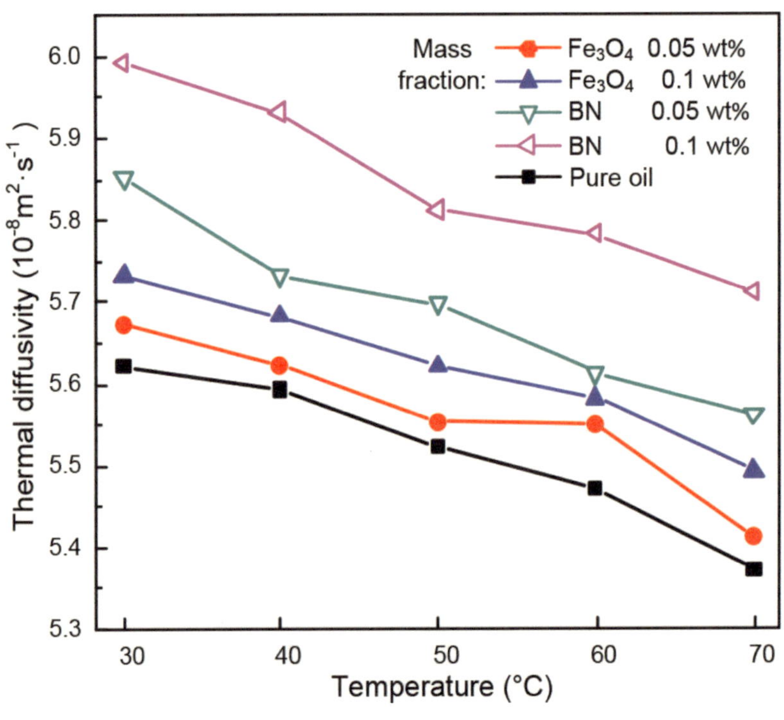

Figure 2. Relation between the thermal diffusivity and the temperature with the different mass fractions.

indicating slight influence of the nanoparticles on temperature-dependent thermal diffusivity in the range of present concentrations.

The improvement in thermal diffusivity is considered to be due to the enhanced thermal conductivity as shown in **Figure 1** which varies with the mass fraction. Moreover, with increasing the loading of nanoparticles, the specific heat decreases as shown in **Figure 3** which leads to an increase in thermal diffusivity as mentioned by Murshed et al. [7]. Furthermore, due to the high thermal conductivity of BN nanoparticles, the thermal diffusivity of the BN-modified oil achieves more obvious enhancement in comparison with the Fe_3O_4-modified oil.

Figure 3 illustrates the effect of nanoparticles on the specific heat with different mass fractions. The specific heat decreases with increasing the mass fraction for each type of the nanooil showing the opposite tendency in thermal conductivity and thermal diffusivity tests. The mechanism for the decline of specific heat when introduced with nanoparticles should attribute to the much smaller specific heat of the nanoparticles compared to the base oil [8]. The result also shows that the oil modified with Fe_3O_4 nanoparticles obtains higher specific heat at the same mass fraction in contrast with the BN-modified oil.

The temperature change of the coil in heating process is given in **Figure 4**. The transformer oil modified with BN and Fe_3O_4 nanoparticles at the mass fraction of 0.1% was applied along with the pure oil. It is observed that the temperature rises lentamente in nano-modified oil and reaches lower maximum temperature indicating the improvement in thermal property

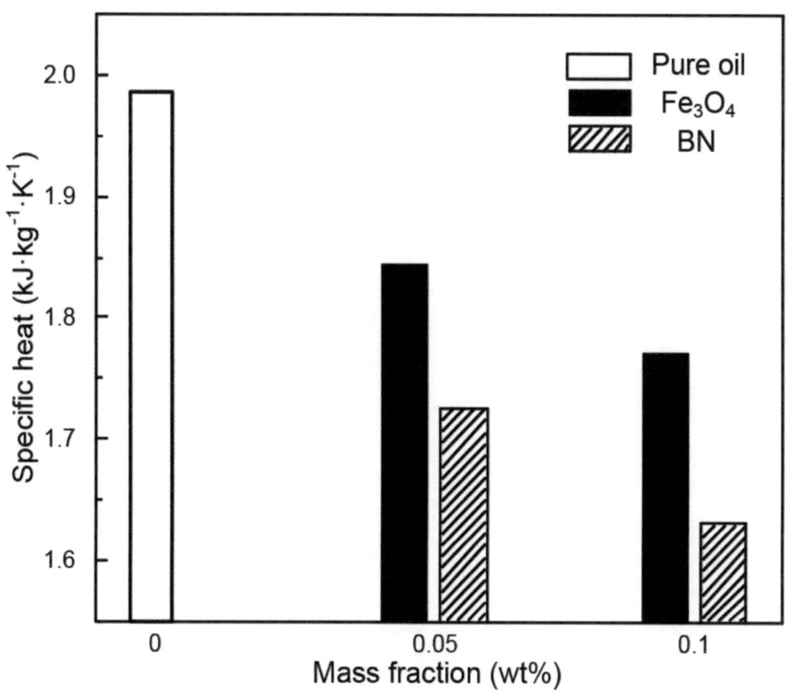

Figure 3. Relation between the specific heat and the mass fraction.

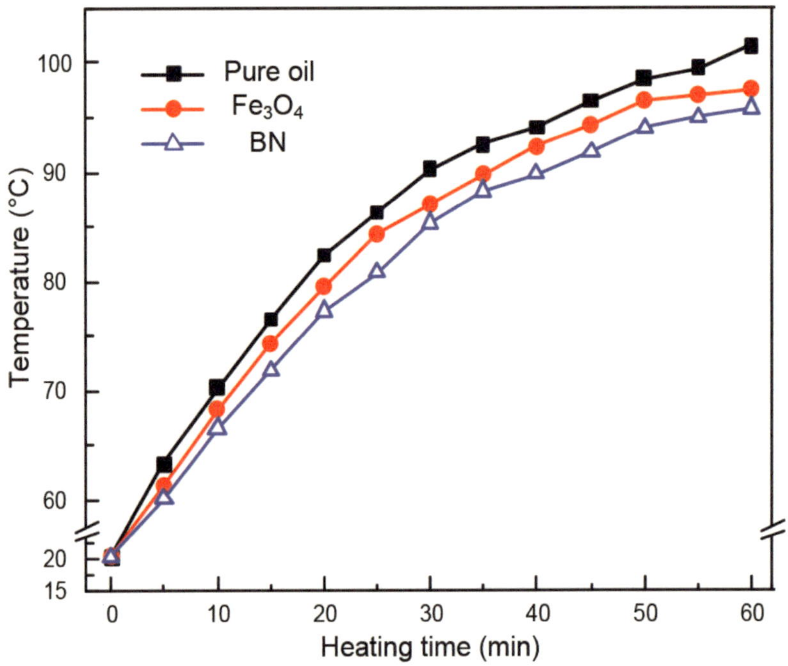

Figure 4. Relation between the heating coil surface temperature and the heating time with the different transformer oil.

with the addition of nanoparticles. For the two types of the nanooil, the temperature of the Fe_3O_4-modified nanooil shows faster increasing tendency compared with the BN-modified nanooil.

Since the nanooil was confirmed to have higher thermal conductivity and thermal diffusivity in previous tests, the heat generated by the coil is expected to be conducted and dissipate to the entire oil in a faster way due to the addition of nanoparticles especially for the BN-modified nanooil. The applied thermal tests confirm the fact that due to the excellent property in thermal conductivity, the transformer oil modified by BN nanoparticles performs superior thermal property compared to the Fe_3O_4-modified oil.

2.3. Effect on dielectric properties

The relative permittivity of the samples is shown in **Figure 5** as a function of the temperature. The oil dispersed with Fe_3O_4 nanoparticles shows higher value than the BN-modified oil at the same temperature and mass fraction. For all the prepared oil samples, the pure oil obtains the lowest value in relative permittivity at the same experimental condition, while the BN-modified oil at the mass fraction of 0.1% achieves nearly the same value as the Fe_3O_4-modified oil at a lower mass fraction of 0.05 wt% in the temperature range. Since the relatively permittivity of transformer oil is lower than the oil-impregnated pressboard, the increase is considered to be beneficial in causing a more uniform electrical field in oil-paper insulation system. As can be seen, the variation of relative permittivity to temperature change for the pure oil performs the same decline tendency as the nano-modified oil.

http://dx.doi.org/10.5772/67432

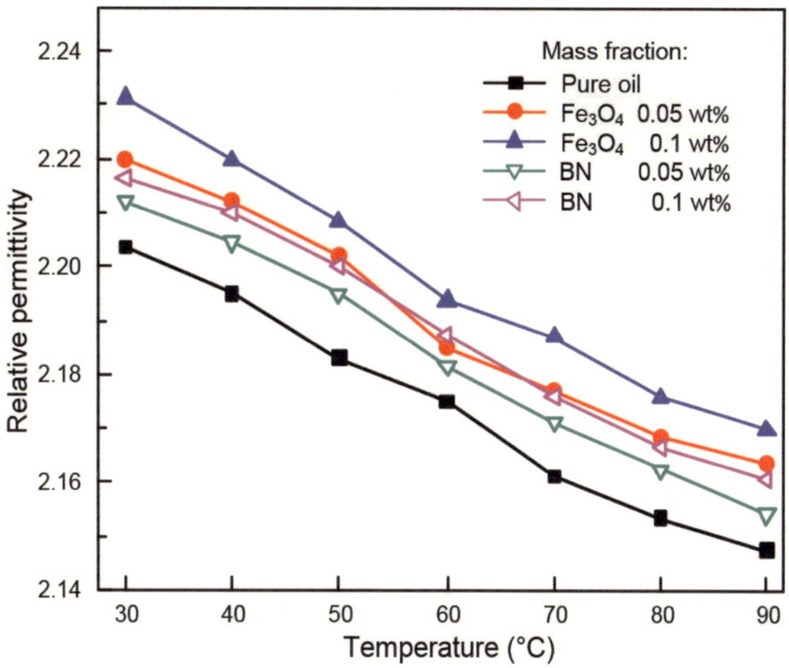

Figure 5. Relation between the relative permittivity and the temperature with the different mass fractions.

The increase of thermal agitation due to temperature rise would make it more difficult for the dipoles to orient and cause the slight decrease in relative permittivity consequently [9]. Due to the high relative permittivity of the additional nanoparticles, the nanooil shows enhanced dielectric property in relative permittivity. Thereby, the higher relative permittivity of Fe_3O_4 nanoparticles causes the higher relative permittivity of the Fe_3O_4-modified nanooil compared to the oil with BN nanoparticles.

Figure 6 shows the measured dissipation factor of the oil at the mass fractions of 0, 0.05 and 0.1% in the temperature range from 30 to 90°C. It is noticed that the dissipation factor of the Fe_3O_4-modified oil is larger than the nonmodified oil and increases with increasing the mass fraction, while the BN-modified oil shows lower value to the pure oil indicating the improved dielectric property with the addition of BN nanoparticles. As can be seen, the dissipation factor of all the oil samples shows the similar tendency in the temperature range. Furthermore, with increasing the mass fraction, the increasing tendency slows down obviously for the BN-modified oil.

The change of dissipation factor to temperature variation is considered to be mainly due to the change of electrical conductivity at constant frequency. Electrical conductivity significantly increases with temperature as a result of both the increasing dissociation of oil molecule and the decreasing oil viscosity leading to the obvious increase in dissipation factor [10].

The electrical resistivity decreases with increasing the temperature and varies with different additional nanoparticles as shown in **Figure 7**. With increasing the temperature, the carrier mobility depending on viscosity increases resulting in the decrease of electrical resistivity

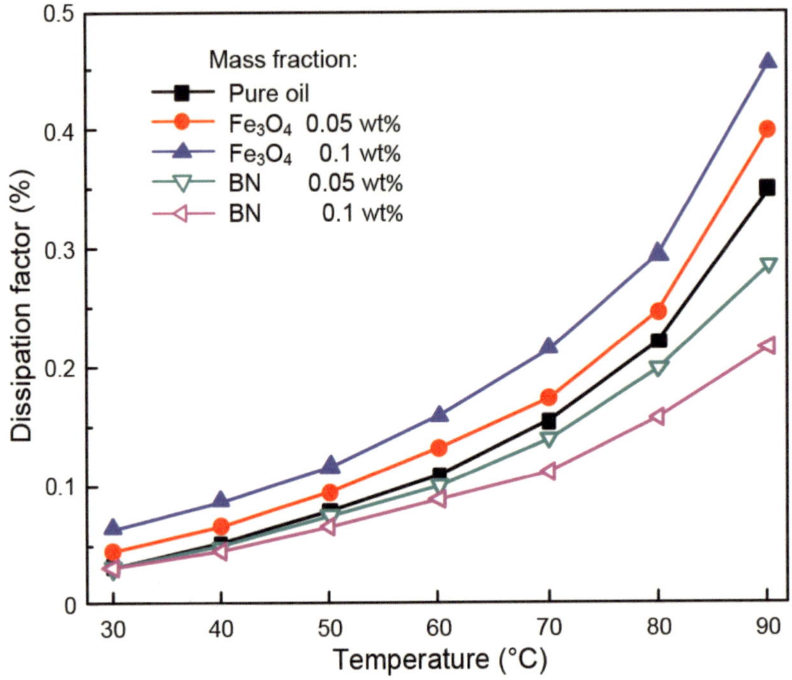

Figure 6. Relation between the dissipation factor and the temperature with the different mass fractions.

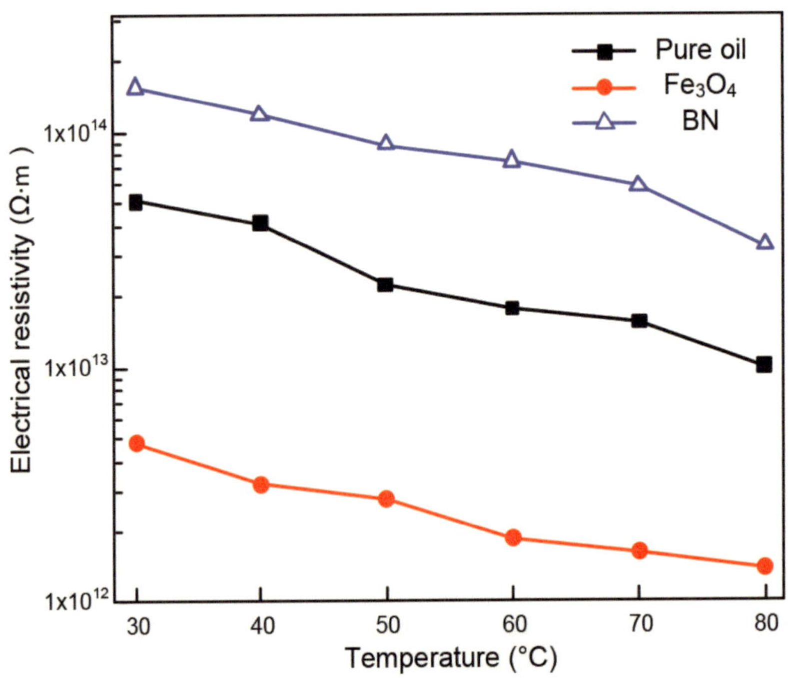

Figure 7. Relation between the electrical resistivity and the temperature with the different transformer oil.

http://dx.doi.org/10.5772/67432

as the characteristic of ionic conduction [11]. It is observed that the transformer oil modified with BN particles shows enhancement compared to the pure oil under the same temperature, while the Fe_3O_4-modified nanooil shows obvious lower value of electrical resistivity confirming the different behaviors between the two types of nanooil in dissipation factor as the result obtained above. The difference in resistivity is regarded as the consequence of the different electrical conductivity of the nanoparticles according to the model for nanooil [12].

2.4. Effect on breakdown strength

The ac breakdown strength varying with temperature and mass fraction is illustrated in **Figure 8**. The ac breakdown strength increases with increasing temperature and follows the similar tendency for all the oil samples. It is observed that the breakdown strength of the nanooil is higher than that of the pure oil in the range of temperature. Moreover, the Fe_3O_4-modified oil reaches even higher value than the BN-modified oil at the same condition. With the increase of mass fraction, the breakdown strength of nanooil rises obviously for both types of the nanooil. As shown in **Figure 9**, the dc breakdown strength of the specimens performs the similar tendency as the nanooil in ac breakdown strength test.

According to the multicore model by Tanaka, the nanoparticles create an interfacial region, which contributes to the improvement in the dielectric strength [13, 14]. It is considered that

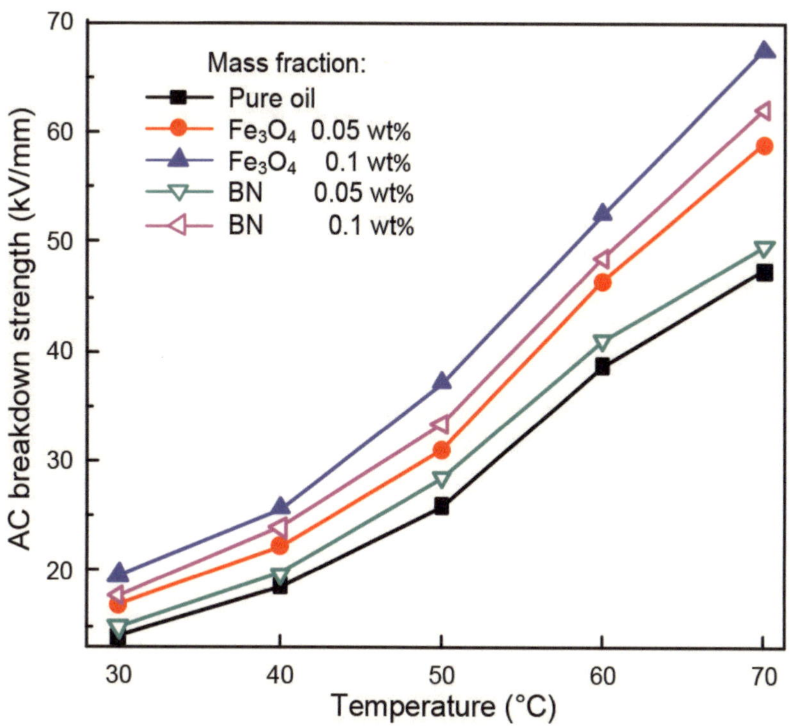

Figure 8. Relation between the ac breakdown strength and the temperature with the different mass fractions.

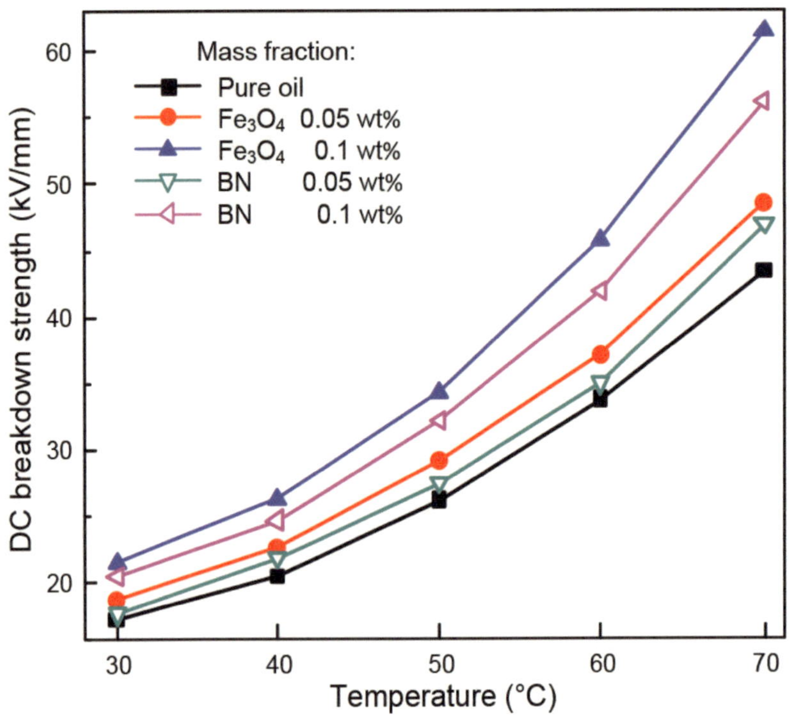

Figure 9. Relation between the dc breakdown strength and the temperature with the different mass fractions.

the percentage saturation dominates in the variation of breakdown strength during the range of temperature [15]. As temperature rises up, the absolute moisture content shows no obvious improvement, while the water solubility increases with temperature. Thereby, the breakdown strength shows the similar increasing tendency with the rise of temperature. Furthermore, Hwang et al found that the charge relaxation time constant can indicate the influence of nano-filter on the electrodynamic process in streamer propagation and explain the obvious enhancement in electric property of the nanooil [11]. Therefore, the different relaxation time constant of the nanoparticles can be explained for the different behaviors in breakdown strength.

Kao suggested a bubble theory for dielectric breakdown in liquid that the breakdown is initiated by bubbles which elongate and cause the formation of conducting channel in liquid gap and eventually result in the occurrence of breakdown due to electrons accelerating [16]. Furthermore, it is considered that the Joule heat generated at the electrodes due to the injection of carries may cause the initial formation of the bubbles [17]. It can be concluded that in the nanooil with enhanced thermal property, the faster heat dissipation to the vicinity results in the inhibition of bubble formation. As can be seen from **Figures 8** and **9**, the nanooil modified with BN nanoparticles shows more obvious improvement compared to the Fe_3O_4-modified oil when the mass fraction increases from 0.05 to 0.1% at the same temperature, which should attribute to the considerable enhancement in thermal property of the BN-modified oil when increase the mass fraction.

The difference in density due to temperature could lead to buoyancy-driven natural convection and cause heat dissipation. Moreover, the viscosity of the liquid shows increase when introduced with nanoparticles [17, 18] and hence may affect natural convection as reported [19]. The change in density and convection could lead to influence on fluid circulation and thus cause effect on the breakdown strength as the flow rate is considered as an important factor on breakdown behavior [15]. However, due to the slight filler loading, no significant increase of the viscosity is expected in the nanooil especially at high temperature. Thereby, it can be concluded that the influence of nanoparticles on convection and the consequent fluid circulation performs no obvious function on the dielectric breakdown strength [20]. Through the analysis and comparison above, it is expected that besides the effect of relaxation time constant, the improved thermal property functions as a considerable factor in the remarkable enhancement of the breakdown strength.

3. Dynamic behavior of surface charge on double-layer oil-paper insulation under pulse voltage

3.1. Experiments

In the research, the paper used in the experiment is produced by Nantong Zhongling Co. Ltd. The oil in the experiment was 25# and produced by Kunlun Energy. Insulation paper with the thickness of 0.13 mm was chosen and cut into the size of 70 mm × 50 mm. The oil and paper were dried at 373.15 K for 24 h first. Then, the paper was immersed into the oil and continued to be dried at the same temperature for another 12 h [22–24]. Finally, the oil-paper composite was placed in the vacuum oven.

3.2. Effect of voltage amplitudes

To show variation characteristics of the surface potential under different voltage amplitudes more clearly, the potential data from 0 to 1200 s are chosen in **Figure 10**, in the partly-enlarged graph, while the trend with the longer time from 0 to 30,000 s is shown in smaller graph. From the two figures, the potential decreases with the increase of time. The decay is fast at the initial period and then becomes slower with the time. The trapped charges in shallow traps are easy to escape, causing the rapid decay at the initial time. With increasing the time, the detrapping of charges in deep traps occurs, and thus, the decrease of surface potential slows down. The lateral charge spreads along the possible routes for charge decay. It either transports through the bulk or recombines with the opposite polar ions in air. The charge is initially transferred from the surface to the bulk in the electric field, then it is trapped and detrapped, and after these processes, the charge finally reaches the grounded electrode [21].

It is also observed in **Figure 10** that as the voltage amplitudes applied in this experiment are significantly different, the difference in initial surface potential is obvious in these curves. It is shown in **Figure 10a** that with different amplitudes, the decay trend is almost the same. However, the initial values of the potential are ~300, ~1550, ~1750 and ~2300. It can be concluded

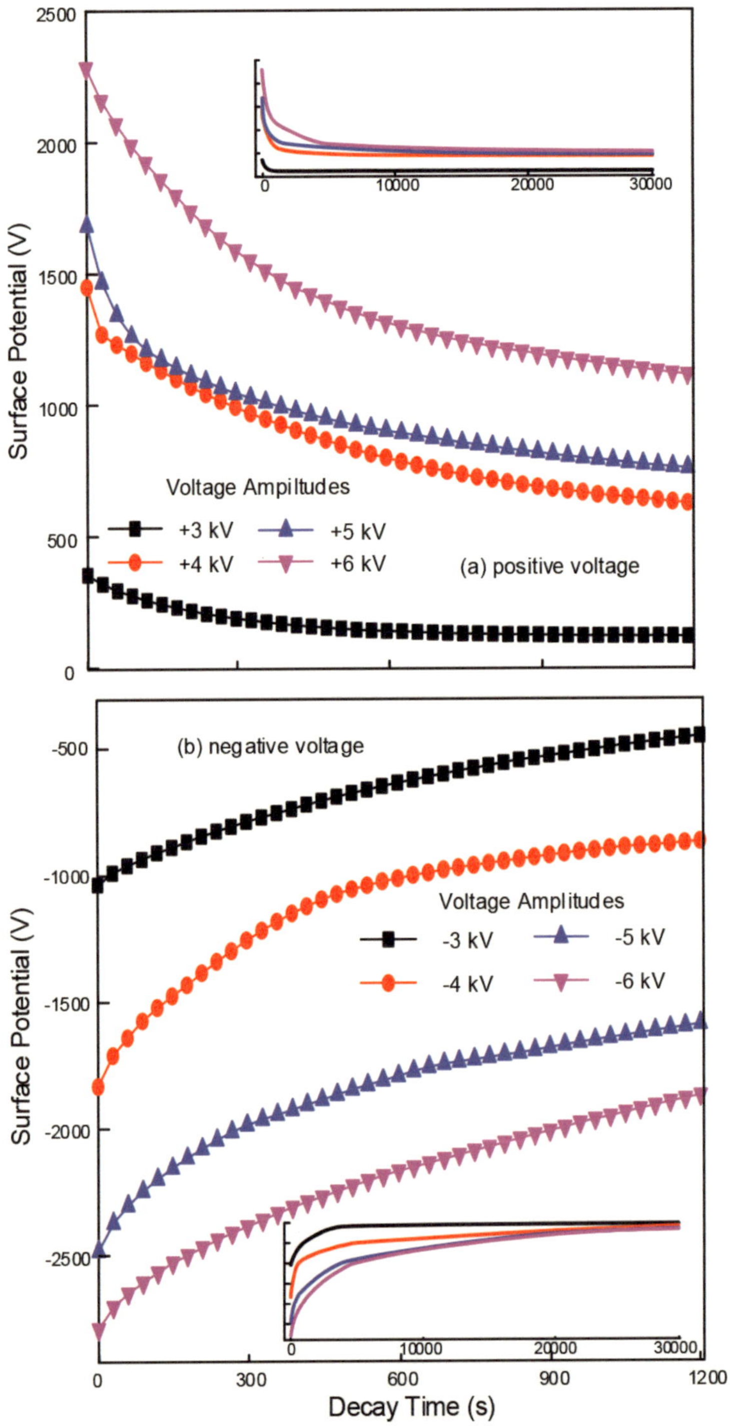

Figure 10. Relationship between the surface potential and the decay time, after a pulse voltage with a frequency of 500 Hz and pulse number of 12,000, with different voltage amplitudes and polarities.

that the increase of voltage amplitudes causes the increase of the surface potential, which indicates that more charge could be injected into paper under the higher pulse voltage [22]. With the time lapse, the surface potential of all four curves decreases. But due to the difference of initial potential value, the curves do not have crossover phenomenon. The surface potential under +3 kV voltage is much lower than that in other cases, which indicates that charge injection needs a threshold value. Only the voltage is applied over this value can the charge be injected into the hole. From **Figure 10b**, things are almost the same. Comparing with the positive pulse voltage, the absolute value of the initial surface potential under the negative pulse voltage is higher under the same experimental conditions. It means that the quantity of positive charges accumulated on the surface is less than negative charges, and it also infers that detrapping process of positive charge is easier than that of negative charge [23].

The relationship between the tdV/dt and the decay time is shown in **Figure 11**. A double exponential decay equation has been used to analyze the decay curve of surface potential, and it was assumed that there were two kinds of decay process [24]. For the tdV/dt curves in **Figure 11a**, it can be inferred that the first peak with a smaller characteristic time corresponds to a faster decay process that the charges escape from the shallow trap, while the second peak with a larger characteristic time is related to a slower decay process that the charges run away from the deep trap. It is clearly showed that the amplitudes of the maximum peak value of +3, +4, +5, and +6 kV are ~80, ~350, ~420, and ~470. Also the corresponding characteristic time changes from 300 to 8000 s, expressing that the peak value and the characteristic time increase in the order of +3<+4<+5<+6 kV. However, in **Figure 11b**, things are different. For the negative pulse voltage, all the four curves have only one peak, indicating that there is only one changing peak of the charge in the corresponding trap depth. When the negative voltage is applied in the paper, a compound peak occurs in the curve.

Furthermore, the negative charge is hard to escape from the traps because it is captured by oil-paper more easily. While the peak values of the curves with the voltage amplitudes of −3, −4, −5 and −6 kV are ~300, ~370, ~700 and ~900, respectively. It is the same with **Figure 11a**, which indicates that with increasing the voltage amplitudes, the peak value of the curves grows. But the characteristic time increases in the order −4<−3<−6<−5 kV. The results show that the duration of the charge decay process with different samples is effected by the voltage amplitudes, but the negative voltage plays a leading role in the characteristic time.

3.3. Effect of pulse frequency

In order to realize how the pulse frequency affects the oil-paper insulation, the relationship between the initial value of surface potential and the pulse frequency is shown in **Figure 12**. Actually, the surface potential of all samples declines quickly in the initial decay period, and then it decreases a little slower than that in the initial time. Moreover, from **Figure 12a**, the initial surface potential in 500, 750 and 1000 Hz with the pulse number of 12,000 is 1565, 1520 and 1324, respectively, which indicates that the initial value of the three curves is very close, while with pulse frequency increasing, the initial becomes a little bit smaller. It indicated that the frequency will slightly affect charge accumulation. **Figure 12b** also corresponds to the same rules, but the absolute value of the surface potential is a bit larger than that in **Figure 12a**.

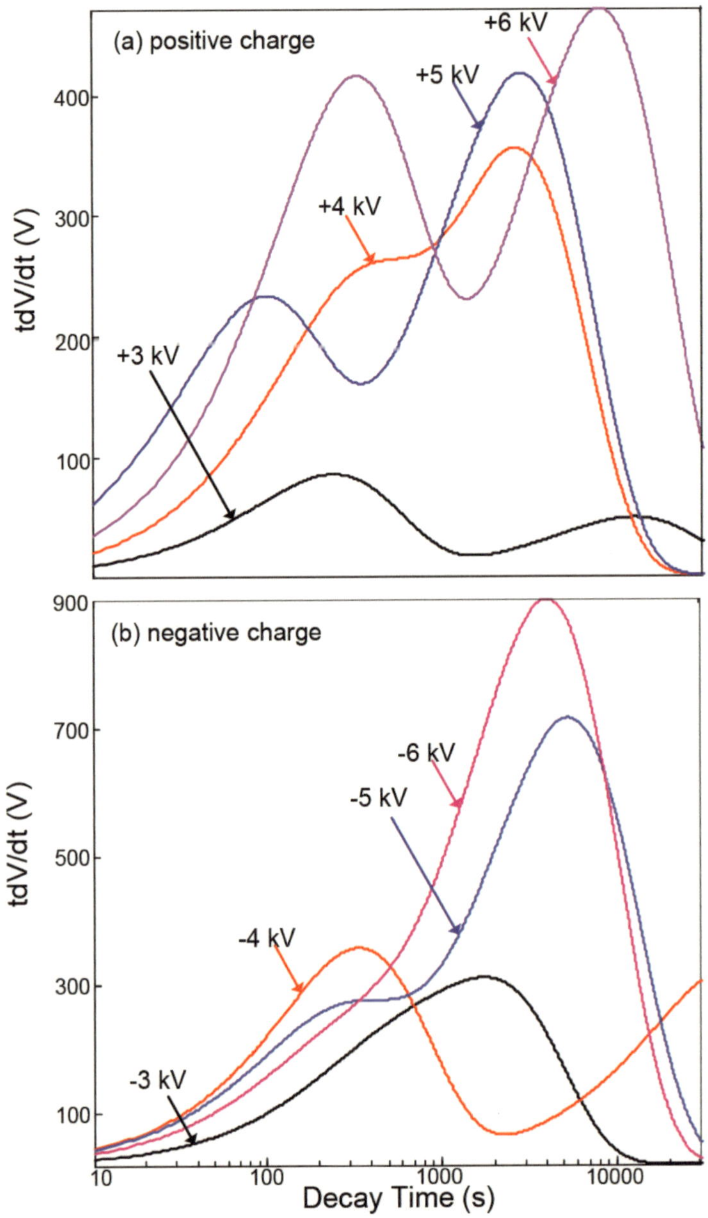

Figure 11. Relationship between the *tdV/dt* and the decay time, after a pulse voltage with a frequency of 500 Hz and pulse number of 12,000, with different voltage amplitudes.

The relationship between the decay rate and the pulse frequency is shown in **Figure 13**. From **Figure 13a**, with the frequency of 500, 750 and 1000 Hz and the pulse number of 12000, the decay rates are 38.2, 44.6 and 52.2%, indicating that with the increase of pulse frequency, the decay rate grows. When the frequency is high, the charge does not have enough time to enter the shallow traps and the energy of injecting charges per unit time becomes higher, then the surface charge becomes less and declines faster. And during the process of charge injection,

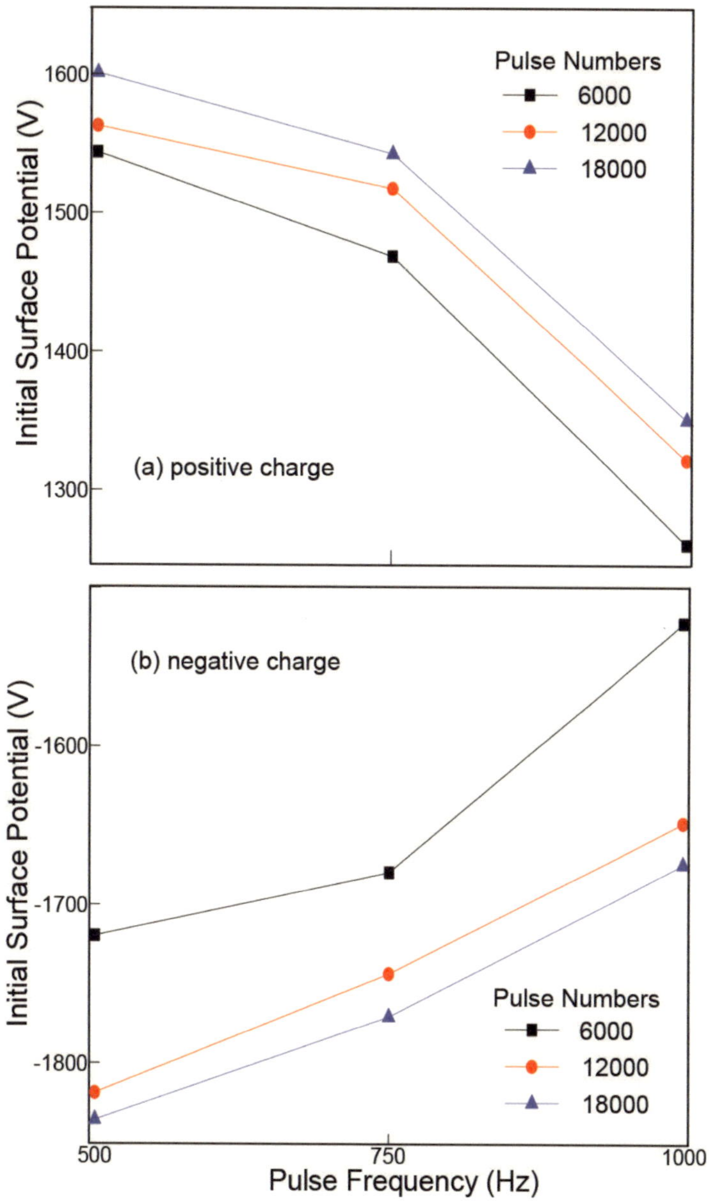

Figure 12. Relationship between the initial surface potential and the pulse frequency, after a pulse voltage with a voltage amplitude of ±4 kV, with different pulse numbers.

the charge injected previously dissipates more because of the higher energy. The decay rate under the negative charging in **Figure 13b** is smaller than that of the positive one, which is due to the detrapping process of negative charge is more difficult than that of positive charge.

Figure 14 displays the relationship between the tdV/dt and the decay time, under different pulse frequency. As in **Figure 14a**, most of the curves also have two peaks for the positive voltage. It is obvious that the maximum peak values of 500, 750 and 1000 Hz are ~290, ~230 and

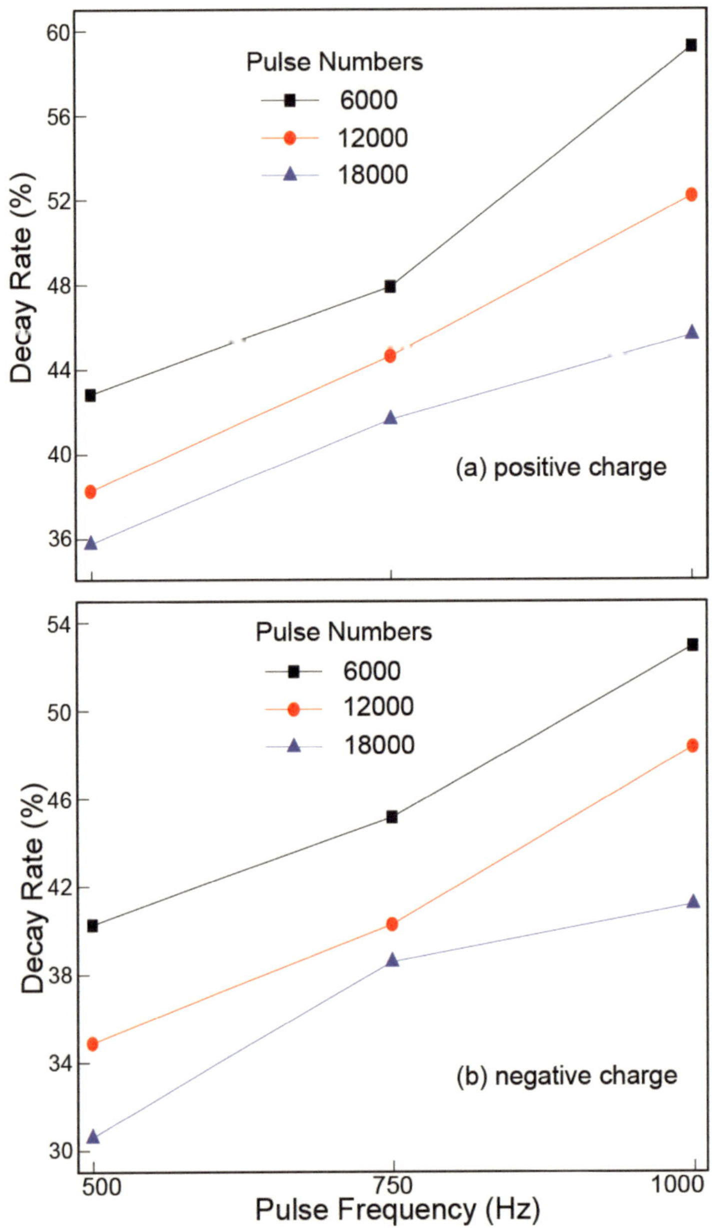

Figure 13. Relationship between the decay rate and the pulse frequency, after a pulse voltage with a voltage amplitude of ±4 kV, with different pulse numbers.

~190, respectively, indicating that the peak value decreases as the frequency grows. The first characteristic time changes from 100 to 1000 s, showing that the decay time decreases with the increase of pulse frequency. In this case, for the sample under the pulse voltage of 1000 Hz, the peak value is the smallest and the characteristic time is the shortest. So the phenomenon in **Figure 15a** accords with that in **Figure 13a**. The similarity is in the case with the negative

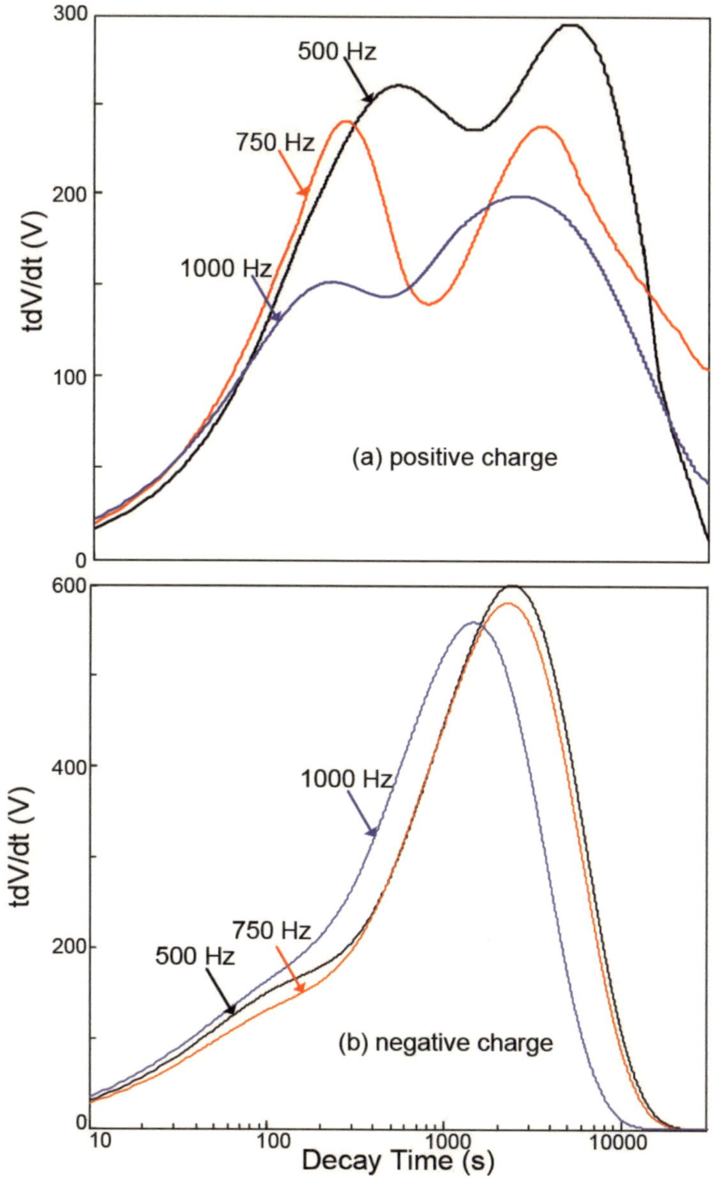

Figure 14. Relationship between the tdV/dt and the decay time, after a pulse voltage with a voltage amplitude of ±4 kV and pulse number of 12,000, with different pulse frequencies.

voltage in **Figure 14b**. For the negative voltage, there is only one peak in every three curves, indicating that there is only one changing peak of the charge in the corresponding trap depth. Also, the peak values of the curves with the frequency of 500, 750 and 1000 Hz are ~600, ~580 and ~550, respectively, which reveals that the increase of the frequency causes the tdV/dt to decrease. The characteristic times decrease with the growth of frequency. It is similar to the situation in **Figure 14a**.

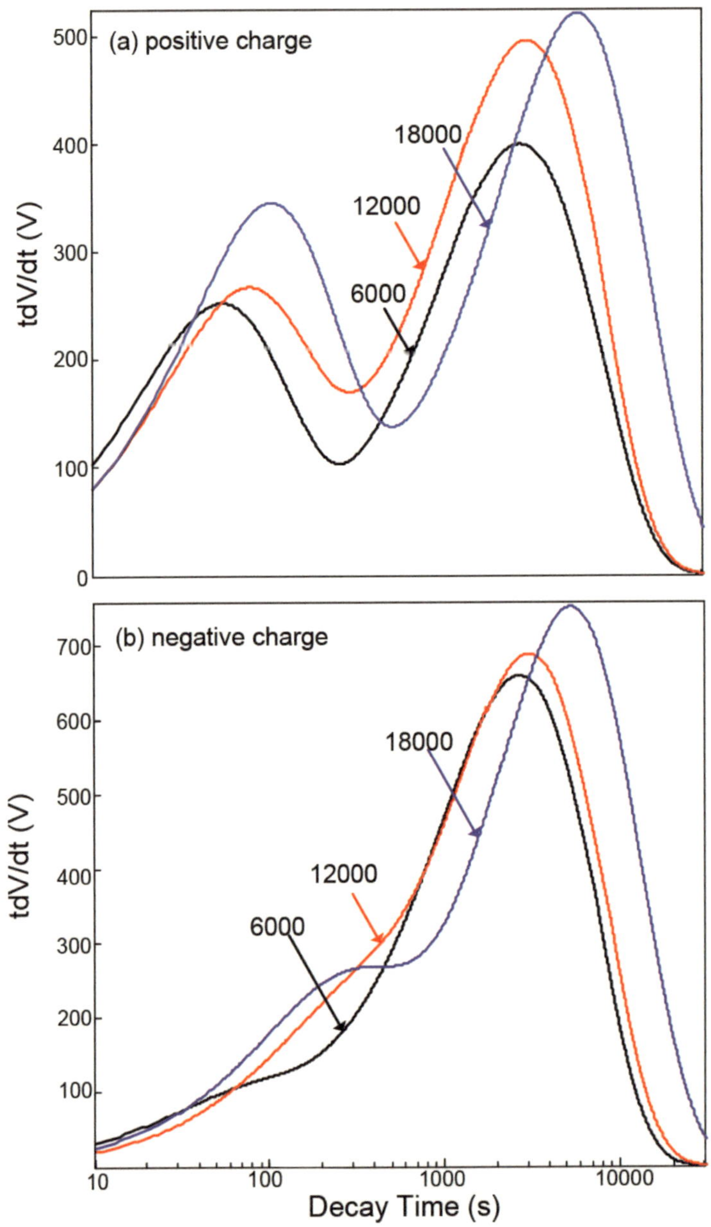

Figure 15. Relationship between the *tdV/dt* and the decay time, after a pulse voltage with a voltage amplitude of ±5 kV and pulse frequency of 750 Hz, with different pulse numbers.

3.4. Effect of pulse numbers

Figure 12 also reflects the relationship between the initial surface potential and the pulse numbers under ±4 kV pulse voltage. It is shown in **Figure 12a** that the initial surface potentials under the pulse numbers of 6000, 12,000 and 18,000 with the frequency of 750 Hz are 1471, 1520 and 1545, respectively, showing that with the increase of pulse numbers, the initial value gets bigger. Actually, when the pulse voltage is applied to the electrode, some ions that are produced by the previous pulse would still exist until the next one comes, which is called the accumulation effect [11]. Therefore, more charges would be easily injected into the surface under more pulses. Also, with increasing pulse numbers, more charges tend to accumulate in the deep traps. **Figure 13a** shows the relationship between the decay rate and the pulse numbers. The decay rate with the pulse numbers of 6000, 12,000 and 18,000 under 750 Hz is 47.9, 44.6 and 41.6%, which expresses that increasing pulse numbers results in a slower charge dissipation. Due to more charges pumped into the deep traps, they would escape from the deep traps and return back to the surface of the paper. So the dissipation of the charges under more pulse numbers would be slowly. **Figures 12b** and **13b** show the similar trend that the absolute value of the potential under the negative pulse voltage is bigger than that under the positive one, while the decay rate is smaller, indicating that the negative charge accumulates more and dissipates more slowly.

Figure 15 shows the relationship between the tdV/dt and the decay time, under different pulse numbers. For the positive voltage, as in **Figure 15a**, most of the curves also have two peaks. And the maximum peak often appears on the second peak. It is obvious that the maximum peak values of 6000, 12,000 and 18,000 pulse numbers are ~400, ~480 and ~530, respectively, which expresses that the peak value increases with the pulse number growing. The first characteristic time changes from 30 to 100 s, indicating that the decay time increases with the growth of pulse numbers. Actually, the second characteristic time also corresponds to the same rule. The peak value is the highest, and the characteristic time is the longest for the paper under the pulse number of 18,000. From **Figure 15b**, the negative voltage is in the similar situation. For the negative voltage, there is only one peak in each curve, indicating that there is only one changing peak of the charge in the corresponding trap depth. Also, the peak values of the curves with the pulse numbers of 6000, 12,000 and 18,000 are ~650, ~680 and ~750, respectively, which reveals that the growth of the pulse number causes the tdV/dt to increase. The characteristic time changes from 2000 to 4000 s, and it increases as the numbers grow.

4. Charge coupling behavior of double-layer oil-paper insulation under DC and pulse voltages

4.1. Experiments

In this research, insulation paper with a thickness of 0.13 mm was chosen and cut to the size of 70 mm × 50 mm. Then, the transformer paper was dried at 373.15 K for 24 h. Later, the paper was immersed in the oil and continued to be dried at the same temperature for another 12 h. Finally, the composite oil-paper was laid into the vacuum oven for 1 day.

4.2. Effect on charge coupling with different voltage waveforms

Figures 16 and **17** show the relationship between the surface potential and the decay time. **Figure 16a** shows the situation where the oil-paper insulation was applied to positive DC and pulse voltages. As the pulse voltage amplitudes applied in this experiment are different, initial

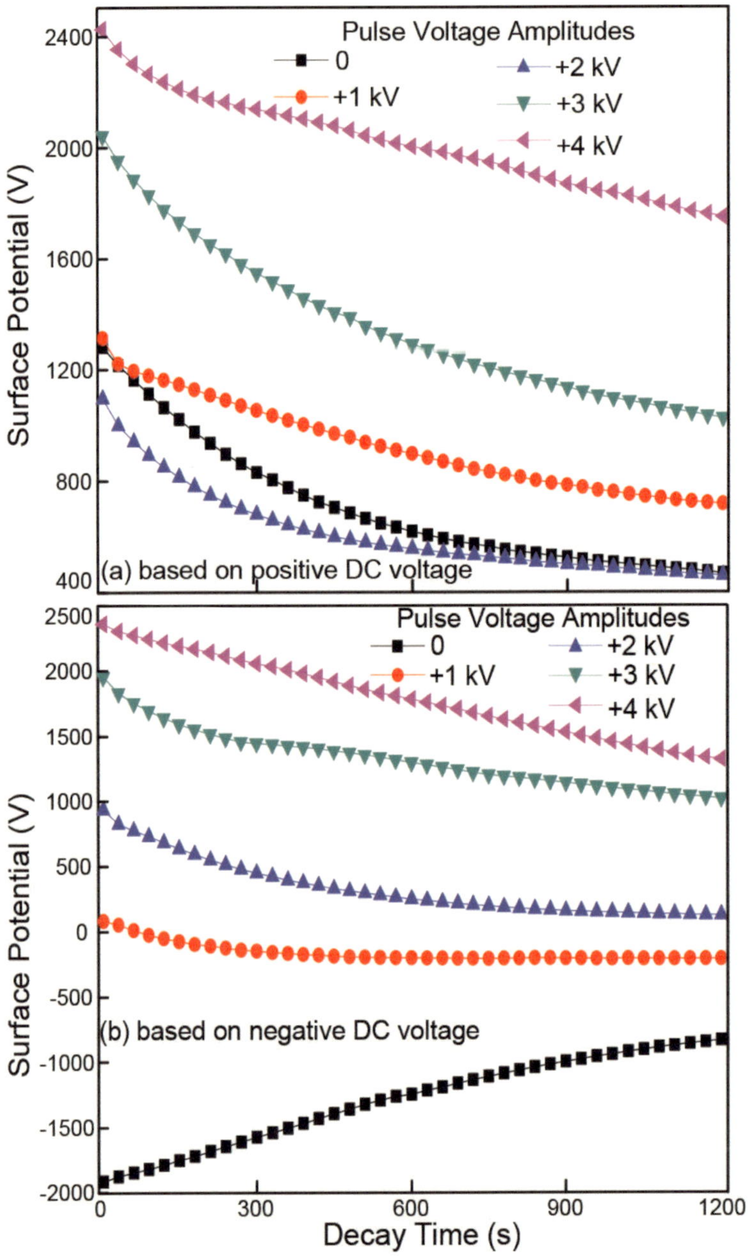

Figure 16. Relationship between the surface potential and the decay time, after an application of DC and positive pulse voltage with a frequency of 500 Hz and pulse numbers of 6000, with different voltage amplitudes.

values of the surface potential are obviously different in these curves. **Figures 16b** and **17a** show that, with the increase of pulse voltage amplitudes, the charge with the same polarity as the pulse voltage increases. It is not the same as the situation in **Figures 16a** and **17b** when compared with the initial values. **Figure 18** represents the rules more obviously. According to

http://dx.doi.org/10.5772/67432

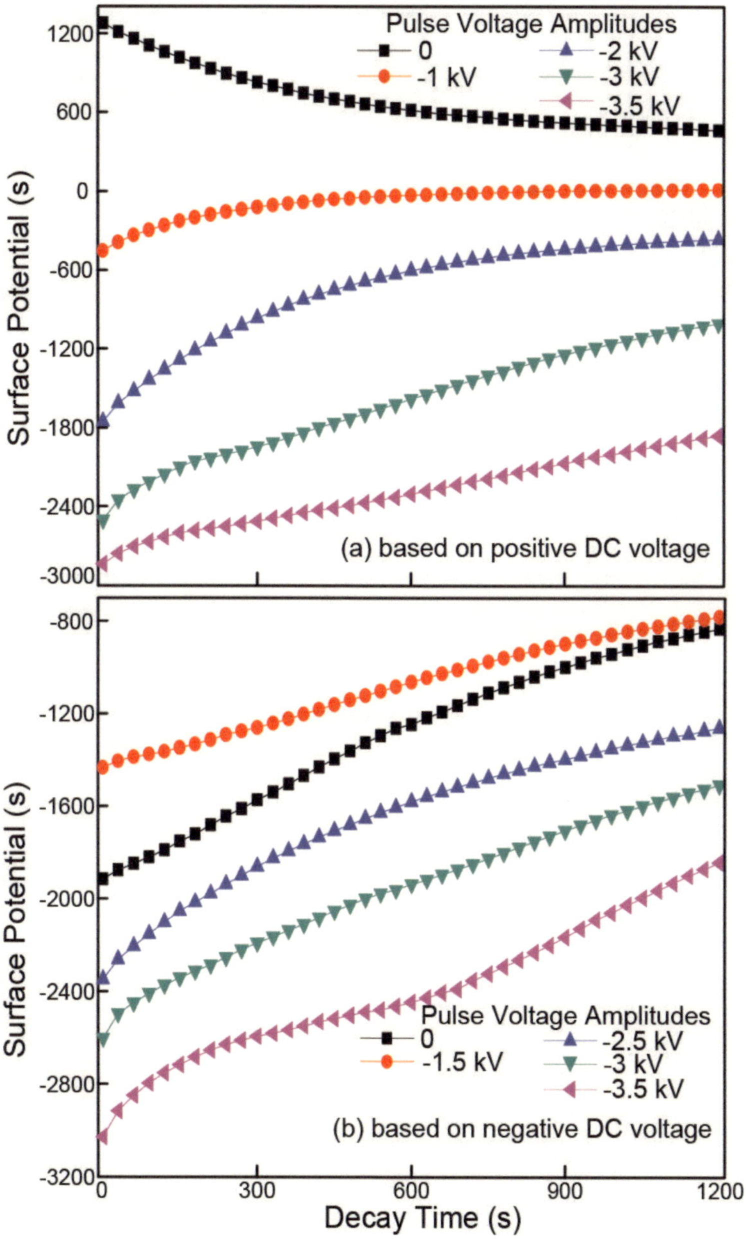

Figure 17. Relationship between the surface potential and the decay time, after an application of DC and negative pulse voltage with frequency of 500 Hz and pulse numbers of 6000, with different voltage amplitudes.

the two figures, when the DC and pulse voltages are in the same polarity, the curves show a clear inflection point. When the polarity becomes the opposite, the curves show a single trend.

Now, taking **Figure 18a** as an example, when the pulse voltage amplitudes are changing from 0 to 1 kV, the initial surface potential maintains a fixed value. As the amplitude gets larger,

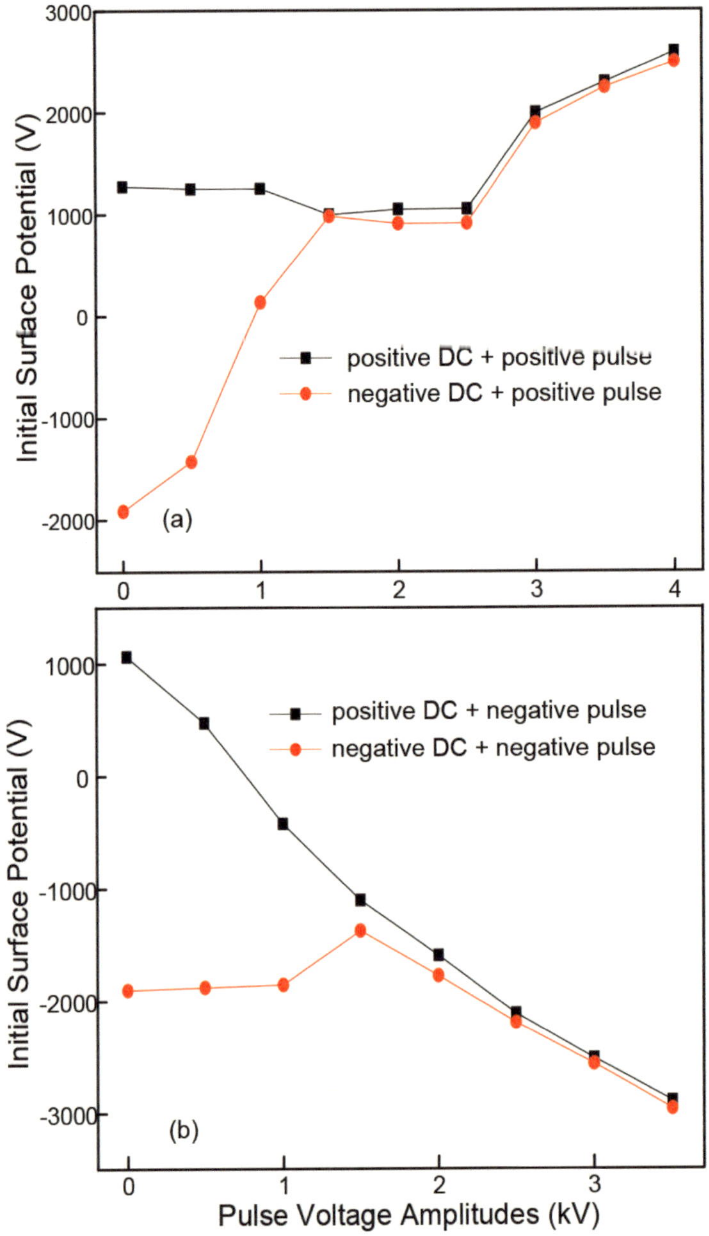

Figure 18. Relationship between the initial surface potential and the pulse voltage amplitudes, after an application of DC and pulse voltage with frequency of 500 Hz and pulse numbers of 6000.

the initial value decreases instead. Furthermore, as the amplitude continues to increase, the initial value grows. In this case, the illustration of charge coupling and accumulation is shown in **Figure 19**. As the DC voltage is applied to the sample, large number of charge are injected from the needle to the sample. Under the same amplitude and duration of DC voltage, the amounts of surface charge are certain. When the pulse voltage amplitude is small enough,

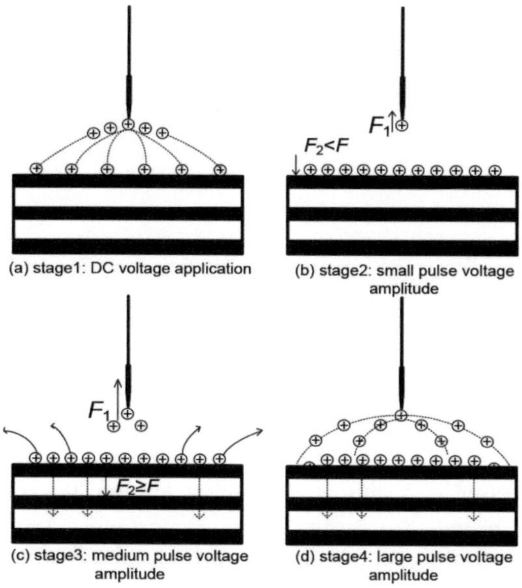

Figure 19. Illustration of surface charge coupling and accumulation, after an application of +3 kV DC and positive pulse voltages in different amplitudes.

which means that the amplitude is no more than 1 kV, as in **Figure 19b**, a small quantity of charge will be injected from the needle.

According to Coulomb's law, due to the repulsive force of the homo-charge, the charge from the needle will no longer be injected into the sample, and of course, the repulsive force F_2 is too small to make the charge from the surface transport through the bulk or recombine with the opposite polar ions in the air ($F_2<F$). So, small pulse voltage amplitude has little effect on initial surface potential. When the amplitude ranges from 1.5 to 2.5 kV, as in **Figure 19c**, the charge from the needle increases with the rise of the amplitude. So, the repulsive force F_1 and F_2 increases. Although the charge from the needle cannot reach the surface of the sample under the repulsive force, some charge from the oil-paper interface will be acted on by enough force to transport through the bulk or neutralize with the opposite polar ions in the air ($F_2 \geq F$). So, the surface charge and the initial surface potential reduce. With the pulse voltage amplitudes increase continuously, as in **Figure 19d**, for more than 2.5 kV, the charge from the needle continues to increase more. At that time, as a result of the strong force and the strong energy of the applied electric field, the pulse voltage will "push" the charge from the needle to the surface of the paper, thus easily raising the surface charge density. So the initial value of the potential will definitely increase, and more charge will be accumulated on the surface of the paper.

When it comes to the negative DC and pulse voltages, the trend is almost the same. The only distinction is that, when the voltages are both positive, as in **Figure 18a**, there will be an almost horizontal line, indicating that when the amplitude of pulse voltage is between 1.5 and 2.5 kV, the initial value of surface potential is almost the same; on the contrary, when the voltages are both negative, as in **Figure 18b**, there is only the highest surface potential, which means that the surface charge is the least at the 1.5 kV pulse voltage. Actually, a positive charge is relatively hard to accumulate on the sample, so homocharge interaction in positive polarity has a relatively complex and long transient process, leading to the lowest saturation values. On the other hand, during the interaction of the negative homo-charge, it is easy to accumulate negative charge.

The illustration of charge coupling and accumulation after a negative DC voltage and a positive pulse voltage applied is shown in **Figure 20**. In the initial stage, due to the negative DC voltage, the negative charge accumulates on the surface of the sample. Then, the positive pulse voltage is applied, and when the voltage amplitude is small, according to **Figure 20b**, on account of Coulomb's law, only small numbers of positive charge are applied to the surface and neutralize with the negative charges. So, there is still some negative surface charge remaining. With the increase of pulse voltage amplitude, more positive charge is pushed to the surface and recombines with the negative. When +1 kV pulse voltage is applied to the sample, from **Figure 20c**, the quantities of positive and negative charge are basically equal. So, the initial surface potential is almost zero. From **Figure 20d**, it can be seen that, as the pulse voltage continues to increase, more positive charge will accumulate on the surface, so the value of the surface potential tends to become positive. When it comes to a positive DC and a negative pulse voltage, the trend is almost the same as that with a negative DC voltage and a positive pulse voltage applied. The only difference is that, when the pulse voltage is positive, as in **Figure 18a**, there will be an almost horizontal line, meaning that when the pulse voltage amplitude is between 1.5 and 2.5 kV, the initial surface potential is almost the same, while when the pulse voltage is negative, as in **Figure 18b**, with the pulse voltage amplitudes increasing, the initial surface potential declines.

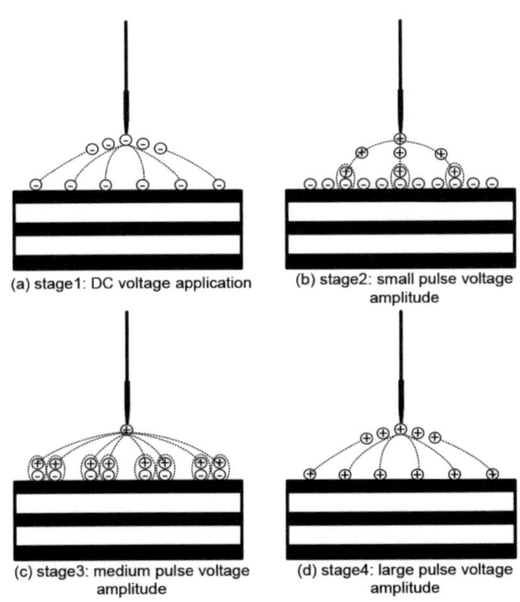

Figure 20. Illustration of surface charge coupling and accumulation, after an application of −3 kV DC and positive pulse voltages in different amplitudes.

Furthermore, from **Figure 18**, it is obvious that, for the same pulse voltage, the absolute initial value of the DC voltage in the same polarity is higher than that in a different polarity, especially for the pulse voltage over 1 kV. Actually, in this situation, with the increase in the amplitude, the gap between the two curves is reduced, but always exists. The reason for this phenomenon is that as DC voltage is in different polarity, the charge polarity is different as well. As the pulse voltage is applied to the sample, it plays a major role in charge accumulation. So the gap decreases with the increase in amplitude. Even so, the charge in the same polarity will increase on the surface, while that in different polarity will first recombine, and then the charge in the same polarity with the pulse voltage will be accumulated. So the absolute value of the DC voltage in the same polarity is ultimately higher than that in different polarity.

Comparing **Figure 18a** and b, the absolute value of the initial surface potential under the negative pulse voltage is larger than that under the positive pulse voltage for a pulse voltage over 1.5 kV. This shows that less positive charge than negative charge is accumulated on the surface, which infers that the detrapping process of the negative charge is more difficult than that of the positive charge [25].

Truly, the pulse voltage surely makes a greater difference than the DC voltage. Through experimental studies, it is obvious that a few pulse voltages over a short time play an important role in charge accumulation. In other words, pulse voltage amplitude indeed greatly affects surface charge. So, pulse voltage is a destructive voltage. Lightning and operating impulse voltages are both factors that may trigger the breakdown of oil-paper insulation. So, it is necessary to ascertain the mechanism of charge coupling and accumulation after DC and pulse voltages, especially the effect of pulse amplitudes.

4.3. Effect on charge coupling with different pulse numbers

In order to understand the effect of pulse numbers on oil-paper insulation, the relationship between the initial surface potential and the pulse numbers is expressed in **Figure 21**. From **Figure 21a**, the initial surface potentials with the pulse numbers of 10, 50, 100, 500, 1000, 3000, 6000, 12,000 and 18,000 for positive DC and pulse voltages are 1872, 1983, 2050, 2112, 2287, 2304, 2342, 2387 and 2390, respectively. Meanwhile, the initial surface potentials with those pulse numbers for a negative DC voltage and a positive pulse voltage are 1787, 1831, 1975, 2083, 2178, 2205, 2242, 2275 and 2346, respectively. The two curves reveal that the initial value of these data shows evident gaps, but the resemblance is that with the increase in pulse numbers, the initial values become a little larger. Meanwhile, as the voltage continues to rise, the initial value increases slowly and becomes gradually saturated. Actually, when the pulse voltage is applied to the electrode, some ions produced by the previous pulse still exist until the arrival of next pulse, and this is called the accumulation effect [26]. Therefore, under more pulses, more charge would be easily injected into the surface, and when the pulse voltage is high enough, there is enough charge accumulated on the surface. Due to the repulsive force, the remaining charge from the needle can no longer be injected to the sample and thus recombines with the opposite polar ions in the air. So the initial value rises with the increase in the pulse number and then comes to saturation. In **Figure 21b**, the two curves show a similar trend, revealing that regardless of the polarities of the DC and pulse voltages, the increase of

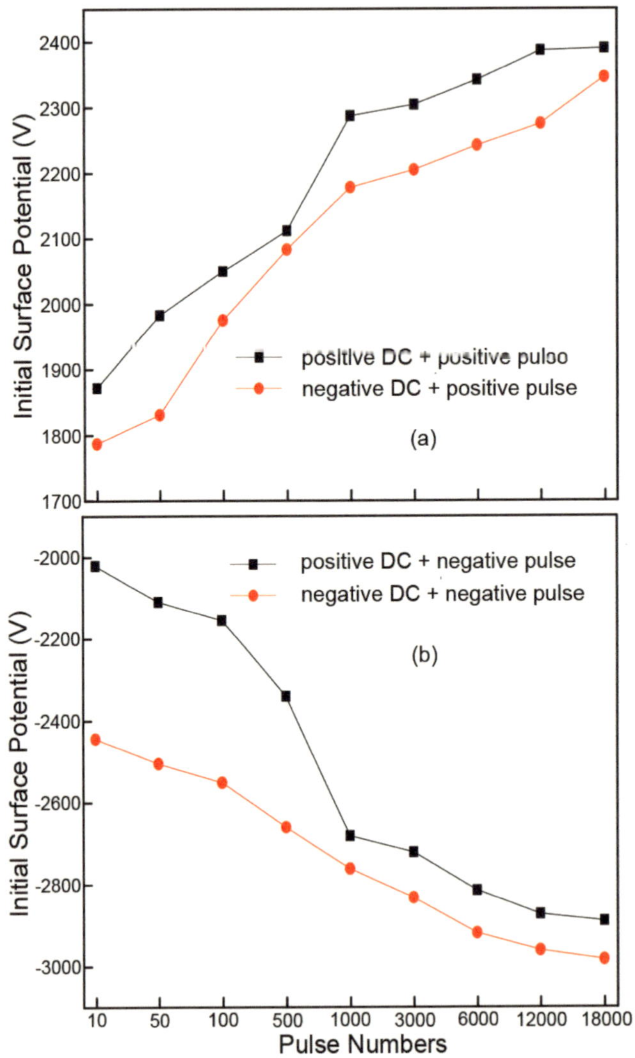

Figure 21. Relationship between the initial surface potential and the pulse numbers, after an application of DC and pulse voltage with voltage amplitude of 3.5 kV, with different voltage polarity.

the pulse number will cause more surface charge to be accumulated, and at last to be saturated. Also, it is evident from these two figures that, when the DC and pulse voltages are in the same polarity, more charge will be accumulated.

The relationship between the decay rate and the pulse number is shown in **Figure 22**, with different voltage polarity. For positive DC and pulse voltages, the decay rates with the pulse numbers of 10, 50, 100, 500, 1000, 3000, 6000, 12,000 and 18,000 are 50.4, 47.4, 44.1, 42.0, 39.0, 36.1, 34.2, 31.1 and 28.7, respectively. Also, for a negative DC voltage and a positive pulse voltage, the decay rates with the corresponding pulse numbers are 57.7, 50.9, 49.7, 45.6, 43.2, 41.5, 37.5, 35.1 and 29.9, respectively. Actually, the other two curves show the same trend, which indicates that with the increase in pulse number, the decay rate decreases, and thus,

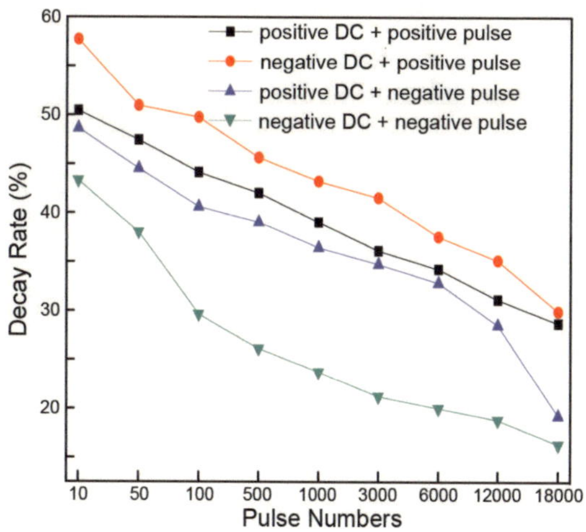

Figure 22. Relationship between the decay rate and the pulse numbers, after an application of DC and pulse voltage with voltage amplitude of 3.5 kV and pulse frequency of 500 Hz, with different voltage polarity.

the charge is dissipated slowly. This phenomenon is due to more charges being pumped into the deep traps, and they will escape from the traps and go back to the surface of the paper. So, the dissipation process of charges under higher pulse numbers would be slow. Also, for the same pulse numbers such as 500, it is easily summarized from the curves that the decay rates of positive DC and pulse voltages, a negative DC voltage and a positive pulse voltage, a positive DC voltage and a negative pulse voltage, and negative DC and pulse voltages are 42.0, 45.6, 39.0 and 26.1%, respectively. Other groups of data show the same trend, revealing that the decay rate decreases with the order as follows: a negative DC voltage and a positive pulse voltage, a positive DC and a positive pulse voltage, a positive DC voltage and a negative pulse voltage, and a negative DC and a negative pulse voltage. There are two reasons for this phenomenon. The main reason is that negative charges are easily accumulated in paper. So, when the negative pulse voltage was applied, the negative charge is easily accumulated and hard to get rid of the traps in the paper. The second reason is that hetero-charges will first recombine with each other, resulting in smaller amounts of surface charge. So, the DC and pulse voltage decay rates in the same polarity are lower than that in different polarity.

The relationship between the tdV/dt and the decay time is shown in **Figure 23**, after a DC voltage and a positive pulse voltage were applied, with different polarity. In these figures, the characteristic time is defined as the time when the curves get to the peak value. For positive DC and pulse voltages, from **Figure 23a**, most of the curves have two peaks, and the maximum peak is usually the second peak. It is obvious that for the first peak values of 10, 50, 100, 500 and 1000, pulses are ~10,000, ~11,000, ~12,000, ~13,000 and ~25,000, respectively, indicating that the peak value increases as the pulse number grows. The first characteristic time changes from 200 to 2000 s, which shows that the decay time increases with the increase in pulse number. In this case, for the sample under the 1000 pulse voltage, the peak value is the largest and the characteristic time is the longest, while for the curve of 1000 pulses, a compound peak appears, showing that there is only one changing peak in the charge in the corresponding trap depth. The existence of the compound peak may be in line with the model

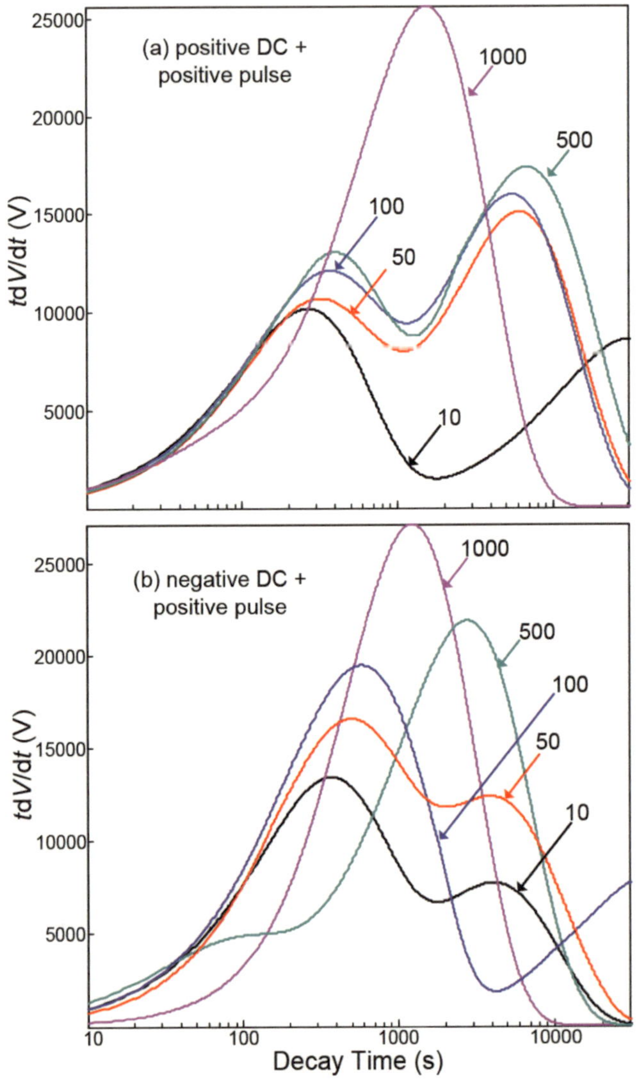

Figure 23. Relationship between the *tdV/dt* and the decay time, after an application of DC and positive pulse voltage with the amplitude of 3.5 kV and pulse frequency of 500 Hz, with different voltage polarity.

proposed by Simmons, which considers the emission current from the closely spaced trapping level [27]. For **Figure 23b**, the initial peak values of the curves with pulses 10, 50, 100, 500 and 1000 are ~13,000, ~17,000, ~20,000, ~22,000 and ~27,000, respectively, also revealing that the growth in the pulse numbers causes the elevation in the *tdV/dt*. The first characteristic time changes from 300 to 3000 s, and it increases in the order 10<50<100<1000<500.

Figure 24 shows the relationship between the *tdV/dt* and the decay time under a DC voltage and a negative pulse voltage. It is clear from **Figure 24a** that, all the five curves only have one peak and for the peak values of 10, 50, 100, 500 and 1000, the pulses are ~17,000, ~24,000, ~26,000, ~30,000 and ~36,000, respectively, indicating that the peak value increases with the

http://dx.doi.org/10.5772/67432

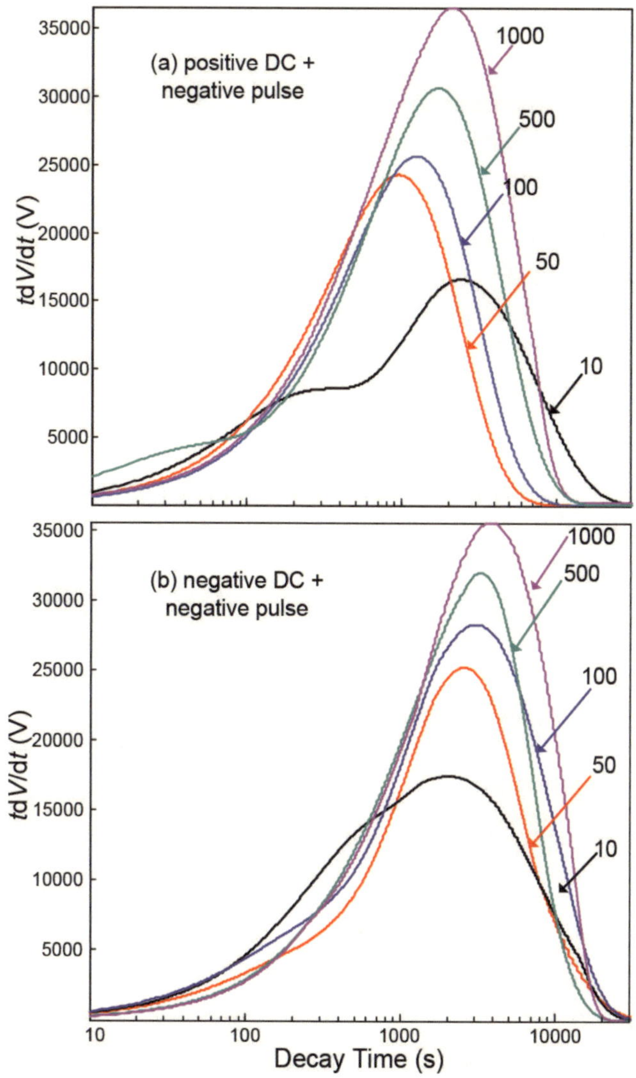

Figure 24. Relationship between the tdV/dt and the decay time, after an application of DC and negative pulse voltage with voltage amplitude of 3.5 kV and pulse frequency of 500 Hz, with different voltage polarity.

pulse number growing. The characteristic time changes from 800 to 2000 s, and it increases in the order 50<100<500<1000<10. From **Figure 24b**, for the negative DC and pulse voltages, there is only one peak in each curve, indicating that there is only one changing peak in the charge in the corresponding trap depth. Also, the peak values of the curves with the corresponding pulse numbers are ~17,000, ~25,000, ~28,000, ~32,000 and ~35,000, respectively, revealing that the growth in the frequency causes the rise in the tdV/dt. The characteristic time changes from 2000 to 4000 s, and it increases as the numbers enlarge. Also, in comparing **Figures 23** with **24**, it is obvious that the peak value of the negative pulse voltage is commonly larger than that of the positive pulse voltage, which probably means that the negative pulse voltage has higher corona discharge energy [28].

5. Conclusion

The effect of BN and Fe_3O_4 nanoparticles on thermal and dielectric characteristics of the transformer oil and the effect of voltage amplitudes, pulse frequencies, numbers and polarities were investigated. This chapter suggested a method to modify the properties of transformer oil, and obtained results illustrate obvious enhancement both in breakdown strength and in thermal properties due to the addition of nanoparticles. From the discussion of the relationship between the surface charge behavior and these key elements, there is better understanding of recognizing the charge coupling dynamics according to the surface potential and the tdV/dt change rule between the paper layers in a multilayer insulation system. The main conclusions can be summarized as follows:

1. The thermal conductivity and thermal diffusivity increase with increasing the mass fraction of the two types of nanooil. The improvement could attribute to the interfacial region due to the additional nanoparticles and the ballistic phonon transport among the nanoparticles. The BN-modified nanooil shows superior thermal property in comparison with Fe_3O_4-modified oil due to the high thermal conductivity of BN nanoparticles.

2. Compared with the pure oil, the relative permittivity of the two types of nanooil is higher in the temperature range due to the high relative permittivity of the nanoparticles. The dissipation factor is lower, and electrical resistivity is higher in BN-modified oil while showing the opposite tendency in the Fe_3O_4-modified oil due to the different dielectric property of the two types of nanoparticles.

3. The breakdown strength of the two types of nanooil shows obvious enhancement with the addition of the nanoparticles which could attribute to the interfacial region created by the nanoparticles and increases with increasing the temperature due to the decrease of relative moisture content. The different improvement to mass fraction for the two types of nanooil indicates that the improved thermal property acts as an important role in the enhancement of breakdown strength according to the bubble theory.

4. With increasing the voltage amplitude, if the polarity of the DC and pulse voltages is same, the absolute value of the initial surface potential decreases initially and then increases, while, if two kinds of voltages are in different polarity, the charge with the same polarity as the pulse voltage increases. Furthermore, for the same pulse voltage, the absolute initial value of the DC voltage in the same polarity is higher than that in different polarity, and the absolute value of the initial surface potential under the negative pulse voltage is higher than that under the positive pulse voltage. So, the pulse voltage plays a more main role than DC voltage in charge accumulation.

5. Taking account of the pulse number effect, it is evident that the initial value rises with the pulse number increasing and then reaches saturation. Considering the decay rate characteristics, with the pulse number growing, the decay rate declines, and thus, the charge dissipates slowly. Also, the decay rate decreases in the following order: a negative DC voltage and a positive pulse voltage, a positive DC and a positive pulse voltage, a positive DC voltage and a negative pulse voltage and a negative DC and a negative pulse voltage.

http://dx.doi.org/10.5772/67432

Author details

Boxue Du

Address all correspondence to: duboxue@tju.edu.cn

School of Electrical Engineering and Automation, Tianjin University, Tianjin, China

References

[1] Xue Y, Zhang XP. Reactive power and AC voltage control of LCC HVDC system with controllable capacitors. IEEE Transactions on Power Systems. 2017;**32**(1):753–764.

[2] Qi B, Wei Z, Li CR. Creeping discharge of oil-pressboard insulation in AC-DC composite field: phenomenon and characteristics. IEEE Transactions on Dielectrics and Electrical Insulation. 2016;**23**(1):237–245.

[3] Murdiya F, Hanaoka R, Akiyama H, Miyagi K, Takamoto K, Kano T. Creeping discharge developing on vegetable-based oil/pressboard interface under AC voltage. IEEE Transactions on Dielectrics and Electrical Insulation. 2014;**21**(5):2102–2110.

[4] IEEE Guide for Loading Mineral-Oil-Immersed Transformers, IEEE Standard C57.91, 1995.

[5] Kole M, Dey TK. Role of interfacial layer and clustering on the effective thermal conductivity of CuO-gear oil nanofluids. Experimental Thermal and Fluid Science. 2011;**35**(7): 1490–1495.

[6] Keblinski P, Phillpot SR, Choi SUS, Eastman JA. Mechanisms of heat flow in suspensions of nano-sized particles (nanofluids). International Journal of Heat and Mass Transfer. 2001;**45**(4):855–863.

[7] Murshed SMS, Leong KC, Yang C. Determination of the effective thermal diffusivity of nanofluids by the double hot-wire technique. Journal of Physics D. 2006;**39**(24):5316–5322.

[8] Murshed SMS. Simultaneous measurement of thermal conductivity, thermal diffusivity, and specific heat of nanofluids. Heat Transfer Engineering. 2012;**33**(8):722–731.

[9] Rajab A, Sulaeman A, Sudirham S, Suwarno. A comparison of dielectric properties of palm oil with mineral and synthetic types insulating liquid under temperature variation. Institut Teknologi Bandung (ITB). Journal of Engineering and Technological Sciences. 2011;**42**(3):191–208.

[10] Li J, Zhang ZT, Zou P. Preparation of a vegetable oil-based nanofluid and investigation of its breakdown and dielectric properties. IEEE Electrical Insulation Magazine. 2012;**28**(5):43–50.

[11] Hwang JG, Zahn M, O'Sullivan FM, Pettersson LAA, Hjortstam O, Liu RS. Effects of nanoparticle charging on streamer development in transformer oil-based nanofluids. Journal of Applied Physics. 2010;**107**(1):1–17.

[12] J. Miao, M. Dong and L. P. Shen, "A Modified Electrical Conductivity Model for Insulating Oil-Based Nanofluids", IEEE International Conference on Condition Monitoring and Diagnosis (CMD), pp. 1126–1129, 2012.

[13] Tanaka T. Dielectric nanocomposites with insulating properties. IEEE Transactions on Dielectrics and Electrical Insulation. 2005;**12**(2):914–928.

[14] Smith RC, Liang C, Landry M, Nelson JK, Schadler LS. The mechanisms leading to the useful electrical properties of polymer nanodielectrics. IEEE Transactions on Dielectrics and Electrical Insulation. 2008;**15**(1):187–196.

[15] Jayaram S. Effects of thermal and viscous drag forces on AC breakdown characteristics of transformer oil. IEEE Conference on Electrical Insulation and Dielectric Phenomena (CEIDP); 1993. pp. 396–401.

[16] Kao KC. Theory of high-field electric conduction and breakdown in dielectric liquids. IEEE Transactions on Electrical Insulation. 1976;**11**(4):121–128.

[17] Vasheghani M, Marzbanrad E, Zamani C, Aminy M, Raissi B, Ebadzadeh T, Barzegar-Bafrooei H. Effect of Al_2O_3 phases on the enhancement of thermal conductivity and viscosity of nanofluids in engine oil. Heat Transfer Engineering. 2011;**47**(11):1401–1405.

[18] Timofeeva EV, Moravek MR, Singh D. Improving the heat transfer efficiency of synthetic oil with silica nanoparticles. Journal of Colloid and Interface Science. 2011;**364**(1):71–79.

[19] Putra N, Roetzel W, Das SK. Natural convection of nano-fluids. Heat and Mass Transfer. 2003;**39**(8–9):775–784.

[20] Jin HF, Andritsch T, Tsekmes IA, Kochetov R, Morshuis PHF, Smit JJ. Properties of mineral oil based silica nanofluids. IEEE Transactions on Dielectrics and Electrical Insulation. 2014;**21**(3):1100–1108.

[21] Wang FP, Xia ZF, Qiu Xl, Shen J, Zhang XP, An ZL. Piezoelectric properties and charge dynamics in poly(vinylidene fluoride-hexafluoropropylene) copolymer films with different content of HFP. IEEE Transactions on Dielectrics and Electrical Insulation. 2006;**13**(5):1132–1139.

[22] Tamura R, Miura Y, Watanabe T, Iahii T, Yamada N, Nitta T. Static electrification by forced oil flow in large power transformer. IEEE Transactions on Power Apparatus and Systems. 1980;**99**(1):335–343.

[23] Du BX, He ZY, Du Q, Guo YG. Effects of water absorption on surface charge and dielectric breakdown of polyimide/Al_2O_3 nanocomposite films. IEEE Transactions on Dielectrics and Electrical Insulation. 2016;**23**(1):134–141.

[24] Zhuang Y, Chen G, Chappell PH, Rotaru M. Surface potential decay: Effect of different corona charging times. IEEE Conference on Electrical Insulation and Dielectric Phenomena (CEIDP). 2012.620–623.

[25] Du BX, Li XL, Jiang JP. Surface charge accumulation and decay on direct-fluorinated oil-impregnated paper. IEEE Transactions on Dielectrics and Electrical Insulation. 2016;**23**(5):3094–3101.

[26] Zhang C, Shao T, Ma H, Zhang D, Ren C, Yan P, Tarasenko VF, Schamiloglu E. Experimental study on conduction current of positive nanosecond-pulse diffuse discharge at atmospheric pressure. IEEE Transactions on Dielectrics and Electrical Insulation. 2013;**20**(4):1304–1314.

[27] Simmons JG, Taylor GW, Tam MC. Thermally stimulated currents in semiconductors and insulators having arbitrary trap distributions. Physical Review B. 1973;**7**(8):3714–3719.

[28] Llovera P, Molinie P. New methodology for surface potential decay measurements: Application to study charge injection dynamics on polypropylene films. IEEE Transactions on Dielectrics and Electrical Insulation. 2012;**11**(6):1049–1056.

Section 2

Researches on Typical Polymer Materials

Chapter 5

Statistical Analysis of Partial Discharge during Electrical Tree in Silicone Rubber Nanocomposites under Elevated Temperature

Mohd Hafizi Ahmad

Additional information is available at the end of the chapter

http://dx.doi.org/10.5772/67100

Abstract

The electric fields at the cable accessories such as jointing and termination are not uniform due to the nonuniformity structures of the accessories. Thus, it has attracted the formation of the electrical tree inside the cable accessory that is commonly made from silicone rubber. Also, the location of the cable that is exposed to the high temperature level gives severe effect to the electrical performance of the insulation. Recently, the inclusion of nanoparticles into the cable insulation has resulted in a promising outcome by resisting the discharge phenomenon such as treeing. However, the study on partial discharge during electrical trees grown in silicone rubber nanocomposites under elevated temperature is scarce including the statistical analysis of the partial discharge mechanisms. Therefore, this chapter is aiming to analyze the statistical behaviors of partial discharge during electrical tree growth in silicone rubber nanocomposites under the effect of temperature.

Keywords: partial discharge, electrical treeing, nanocomposites, statistics, temperature

1. Introduction

Nanocomposite materials have been found to be more durable than conventional composites when subjected to high-voltage stresses due to superior thermal, mechanical, and electrical properties. Nanocomposites are defined as composites with very small amounts of homogeneously dispersed nanoparticles added to the matrix by several weight percentage (wt%). Also, the nanoparticles added to the matrix are normally less than 10 wt% in quantity. It has been found that the inclusion of nanoparticles in polymeric insulation may effectively reduce the accumulation of space charge, surface tracking, partial discharges (PD), water treeing, and

INTECH
open science | open minds

© 2017 The Author(s). Licensee InTech. This chapter is distributed under the terms of the Creative Commons Attribution License (http://creativecommons.org/licenses/by/3.0), which permits unrestricted use, distribution, and reproduction in any medium, provided the original work is properly cited.

electrical treeing thereby increasing the lifespan of the insulation. Andritsch et al. [1] reported that the space charge density has been reduced by adding the Magnesia (MgO) nanoparticles to the epoxy resin matrix. Also, Bamji et al. [2] reported that the amount of injected charge in polypropylene with 2 wt% and 4 wt% organoclay was less than that found in pure polypropylene after 500 hours of aging.

A study on an epoxy resin-based nanocomposite material found that partial discharge (PD) magnitudes and PD numbers were lower than that for the unfilled epoxy resin [3]. Alumina (Al_2O_3) nanofiller have also been shown to improve the surface discharge degradation resistance of epoxy resin [4]. Guastavino et al. [5] reported that the inclusion of 5 wt% Montmorillonite (MMT) has increased the breakdown time four times longer than the breakdown time of pure low-density polyethylene (LDPE). In another study by Sridhar and Joy Thomas [6], the inclusion of 1 and 3 wt% of silica (SiO_2) nanofillers in polyethylene (PE) has improved the tree growth resistance and increased the electrical tree inception voltage.

The inclusion of small amounts of nanofillers to a polymer matrix has resulted in significant improvement of the electrical properties of the nanocomposites. The nano-sized particles have larger specific surface area per unit volume compared with the conventional microfillers. Also, the inclusion of nanoparticles into the polymer matrix would introduce the region so-called interaction zones at the interface between the nanoparticles and the host polymer. This interaction view has led to the development of many physical models of the interface regions. One of the models is proposed by Tanaka et al. [7] namely multi-core model to account for the observed nanocomposite behavior. Moreover, the methodology used to uniformly disperse the nanoparticles is considered very important by researchers to achieve the best results.

Furthermore, XLPE has been used widely in insulation cable to cover the voltage levels ranged from 10 to 100 kV and thus the involvement of cable joints and terminals could not be neglected. These accessories are considered as weak points that may contain voids, impurities, defects, protrusions, and so on; all of which may induce an electrical tree. One of the accessories is a stress cone that is commonly made from silicone rubber [8]. In addition, the cable operating temperature can reach up to 90°C, but the usual rated temperature is within 50–60°C [9]. Thus, the performance of insulation at high temperature is important from practical viewpoint because the failure of cable joints has relationship with the temperature due to changes of soil temperature and the presence of hotspot caused by aging process [10]. Furthermore, the high temperature affects the growth of electrical tree in Polyethylene (PE), Ethylene-vinyl Acetate (EVA), High Temperature Vulcanization (HTV), and Room Temperature Vulcanization (RTV) silicone rubbers.

However, the effects of temperature on electrical tree and relevant PD in nanocomposite materials are not discussed thoroughly and the publications on the topics are scarce. Thus, the effect of temperature on inception and propagation of electrical trees in silicone rubber-based nanocomposites was examined in this study. This study was performed because it is believed that despite the overwhelming influence of temperature on electrical tree-associated PD occurrences in silicone rubber/nanoclay nanocomposites, only a few studies have been published in relation with statistical analysis thus requiring the statistical analysis of PD during electrical tree growth in nanocomposites. Besides, the temperatures of the experiment varied from 20°C to 60°C, which lies within the rated operating temperature of most insulated power cables; thus emphasizing the importance of this study as most of the cables have been reported to be operated within 50–60°C [9–11]. Subsequently, statistical analyses of PDs during the electrical tree growth under AC applied voltages as well as under the different temperature conditions were being comprehensively characterized in this chapter.

2. Statistical analysis of PD events

The PD data was analyzed statistically in order to interpret the relationship between the statistical approach and the physical parameters of PD events during the electrical tree growth. Thus, the pulse magnitude distributions $H_n(q)$ and pulse count distribution $H_n(\varphi)$ were considered for the analysis. In the case of $H_n(q)$ and $H_n(\varphi)$, both positive and negative discharge characteristics can be determined and analyzed with four statistical moments that were calculated for each of the distributions.

In general, motivating factor that necessitated the use of conventional statistical tools for the PD analysis was based on the study done by Dodd et al. [12] and Gulski [13]. The difference is that different materials are used; while Dodd et al. [12] and Gulski [13] studied PD activities during electrical tree growth in flexible epoxy resin and PE, respectively, but this study is on the PD activities during tree growth in neat silicone rubber and silicone rubber-based nanocomposites. Thus, different materials would be expected to result in different PD characteristics with PD resistance capability that is expected in silicone rubber-based nanocomposites material.

The analysis of the recorded PDs data in this study was accomplished using MATLAB software. In addition, the power that dissipated in the sample during the PD activity was also calculated based on the following formula [14, 15]:

$$P = \frac{1}{\Delta t}\sum_{i=1}^{N} V_i q_i \quad (1)$$

where Δt is acquisition time that equals to one second, V_i is the instantaneous voltage at which the ith PD event occurs with a magnitude, q_i and N are the total number of PD events in the one second of interval.

2.1. Analysis of PD for neat silicone rubber at temperature of 20°C, 40°C, and 60°C

The four statistical moments, such as mean, standard deviation, skewness, and kurtosis, were computed to get the extension analysis of PDs data. The physical parameters such as average phase of occurrence, number of discharge per second, and positive and negative PD amplitudes per second of data record were also computed. The statistical moments of the pulse magnitude distributions, $H_n(q)$ and pulse count distributions, $H_n(\varphi)$ characterization as a function of time of neat silicone rubber tested at temperature of 20°C, 40°C, and 60°C are illustrated in **Figures 1–6**, respectively, whereas the physical parameters as a function of time of neat silicone rubber tested at 20°C, 40°C, and 60°C are depicted in **Figures 7–9** respectively.

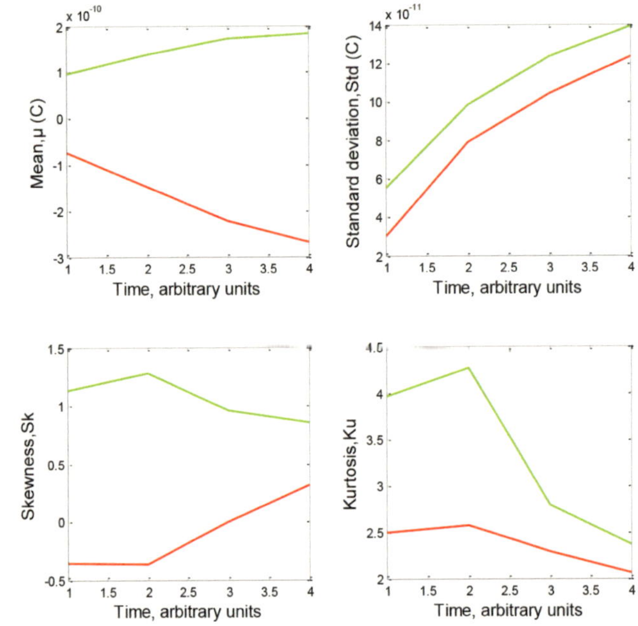

Figure 1. Mean, standard deviation, skewness, and kurtosis of positive and negative PD magnitudes, $H_n(q)$, as a function of time obtained from PDs during tree growth in neat silicone rubber at 20°C. Green solid line = positive PD and red solid line = negative PD.

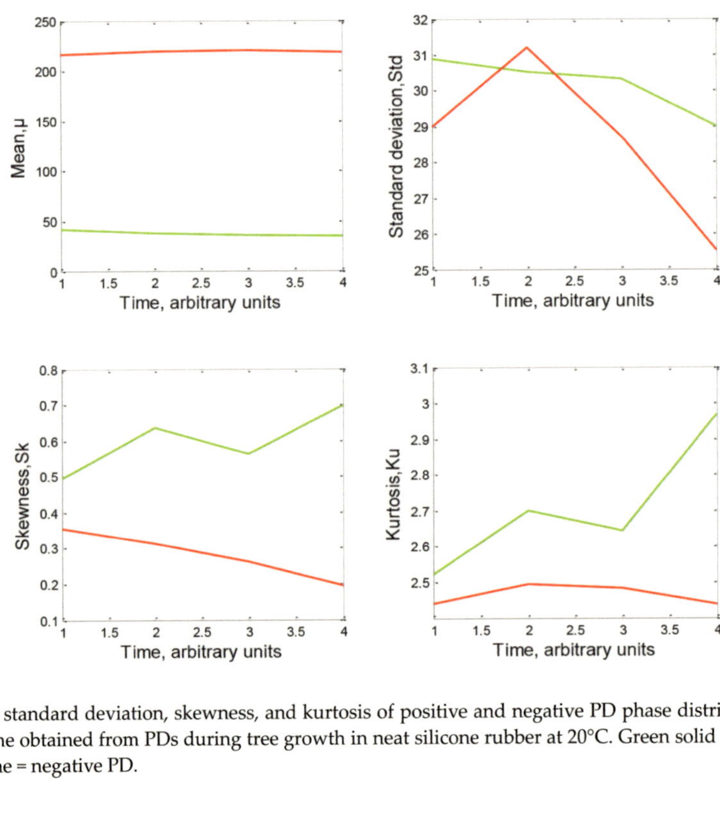

Figure 2. Mean, standard deviation, skewness, and kurtosis of positive and negative PD phase distributions, $H_n(\varphi)$, as a function of time obtained from PDs during tree growth in neat silicone rubber at 20°C. Green solid line = positive PD and red solid line = negative PD.

http://dx.doi.org/10.5772/67100

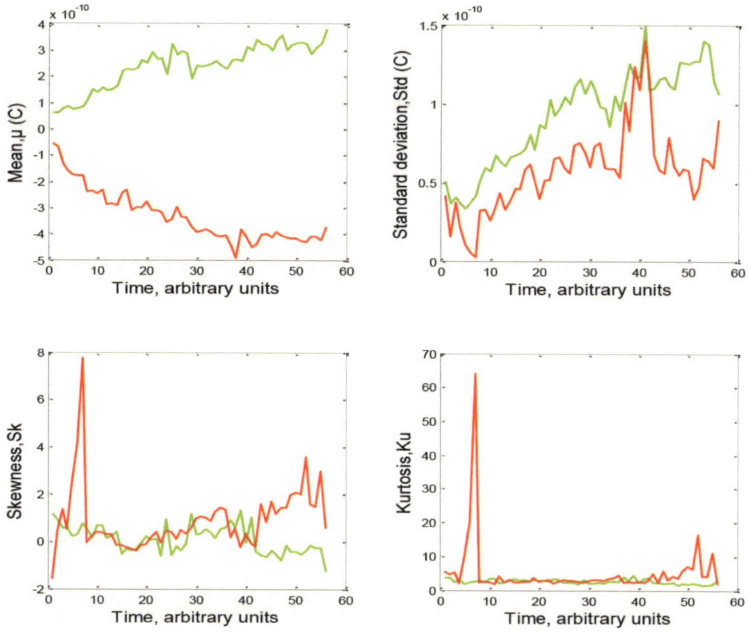

Figure 3. Mean, standard deviation, skewness, and kurtosis of positive and negative PD magnitudes, $H_n(q)$, as a function of time obtained from PDs during tree growth in neat silicone rubber at 40°C. Green solid line = positive PD and red solid line = negative PD.

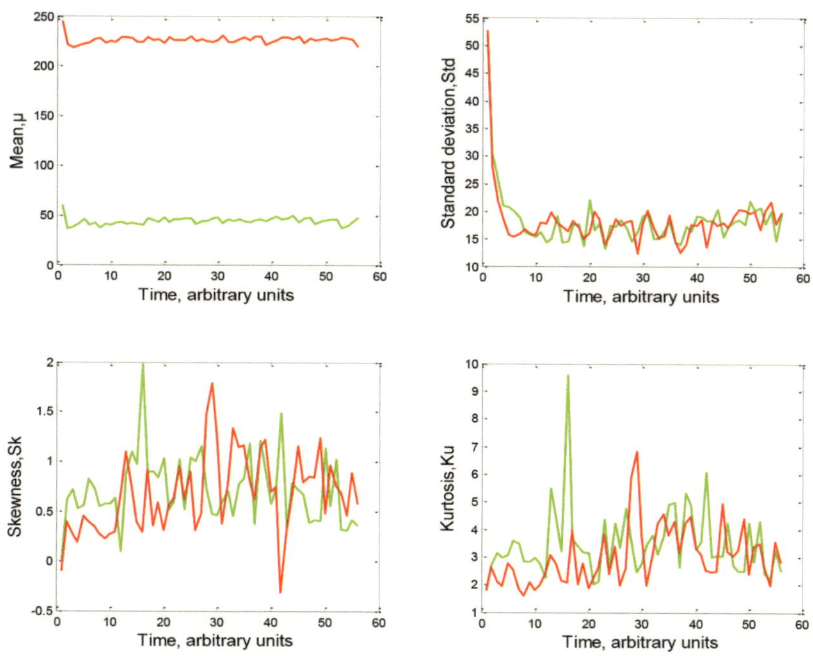

Figure 4. Mean, standard deviation, skewness, and kurtosis of positive and negative PD phase distributions, $H_n(\varphi)$, as a function of time obtained from PDs during tree growth in neat silicone rubber at 40°C. Green solid line = positive PD and red solid line = negative PD.

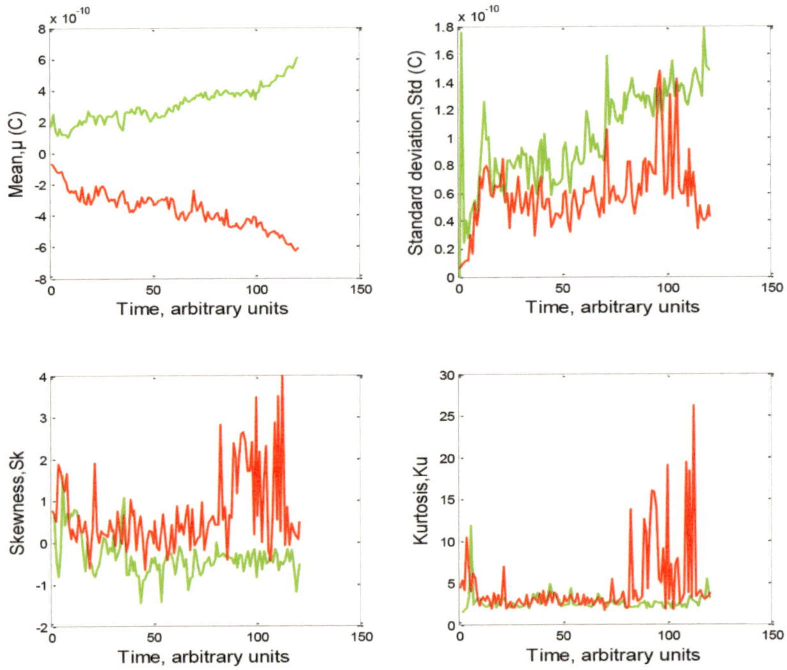

Figure 5. Mean, standard deviation, skewness, and kurtosis of positive and negative PD magnitudes, $H_n(q)$, as a function of time obtained from PDs during tree growth in neat silicone rubber at 60°C. Green solid line = positive PD and red solid line = negative PD.

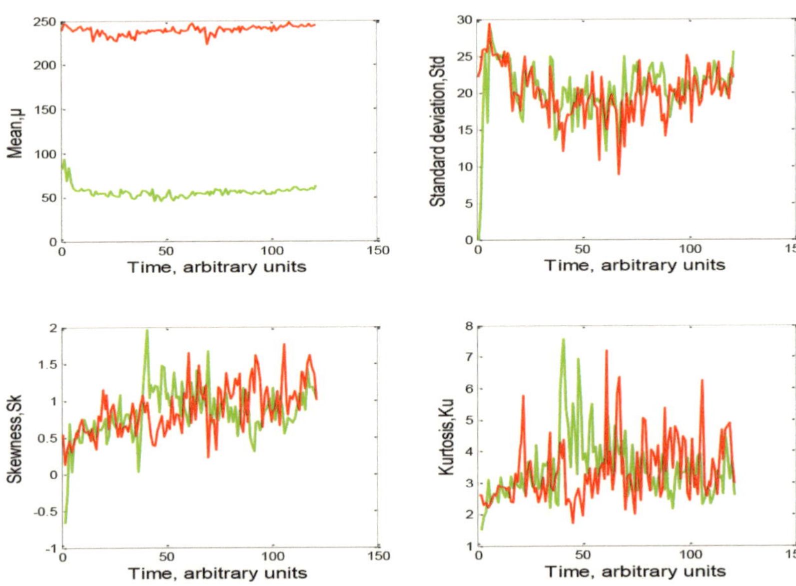

Figure 6. Mean, standard deviation, skewness, and kurtosis of positive and negative PD phase distributions, $H_n(\varphi)$, as a function of time obtained from PDs during tree growth in neat silicone rubber at 60°C. Green solid line = positive PD and red solid line = negative PD.

http://dx.doi.org/10.5772/67100

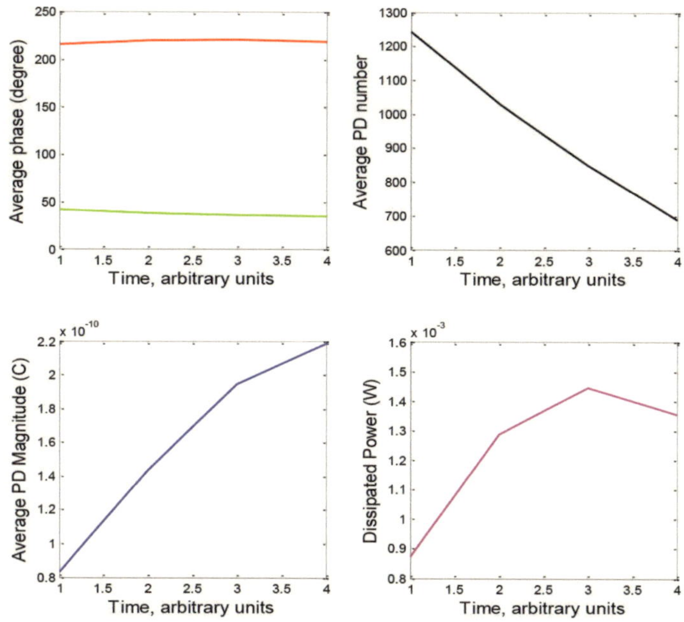

Figure 7. Average phases of occurrence of positive and negative PDs, average number of PDs, average PDs magnitude, and dissipated power as a function of times obtained from PDs during tree growth in neat silicone rubber at 20°C. Green solid line = positive PD and red solid line = negative PD.

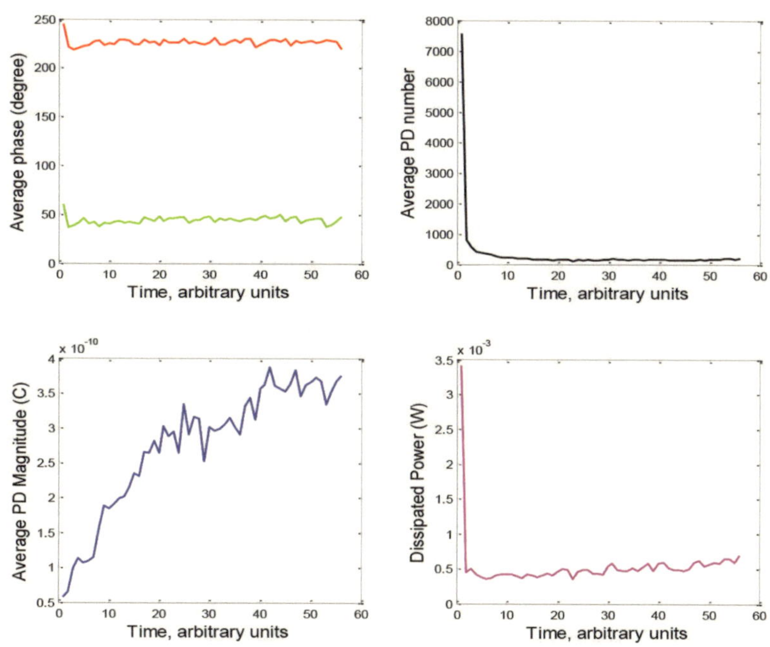

Figure 8. Average phases of occurrence of positive and negative PDs, average number of PDs, average PDs magnitude, and dissipated power as a function of time obtained from PDs during tree growth in neat silicone rubber at 40°C. Green solid line = positive PD and red solid line = negative PD.

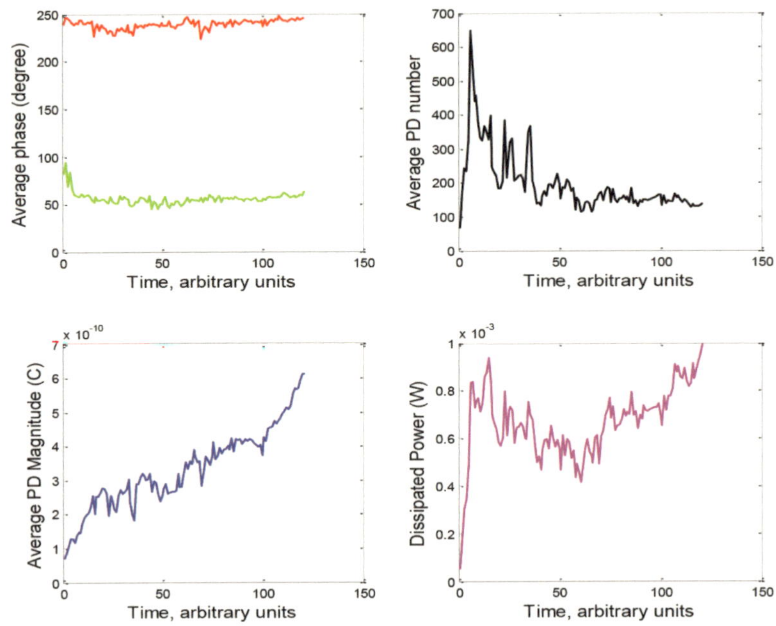

Figure 9. Average phases of occurrence of positive and negative PDs, average number of PDs, average PDs magnitude, and dissipated power as a function of time obtained from PDs during tree growth in neat silicone rubber at 60°C. Green solid line = positive PD and red solid line = negative PD.

PD during electrical tree growth in neat silicone rubber at temperature of 20°C, 40°C, and 60°C with 10 kV_{rms} was subjected to the test samples at relative humidity of 45 ± 10%, and they were also subjected to statistical analyses. According to the published literatures, moisture is one of the factors that would weaken the interface between nanofiller and polymer, leading to the easier splitting of the interface. The moisture could hardly be absorbed into the filler compared with the polymer [16, 17]. In fact, a relative humidity of about 75% around the polymer seems to be necessary for the electrical tree generation [18]. Thus, in this study, we fixed the relative humidity in the range of 35–55%, which is considered low to weaken the interfacial adhesion and influence the growth of treeing. The tree growth time was depicted in arbitrary unit of time. The actual time of the entire tree growth was 80 seconds, thus the unit of time illustrated in **Figure 1** will have to be multiplied by 20 seconds. From **Figure 1**, the mean values of positive PD and negative PD magnitude distributions, $H_n(q)$, were increased with time during the tree growth. The standard deviations of positive and negative PDs were found to increase as well with the increase of growth time. The increase of PD magnitude was related to the increase of tree length in distance, indicating that the PD magnitude and electrical tree length are relative as PDs can propagate in the main body of electrical tree structure [19]. On the other hand, the skewness of positive PD decreases and the skewness of the negative PD increases from negative to positive values over time. Skewness and kurtosis are commonly used to describe the dispersion and shape of PD distribution. The positive skewness indicates the higher PD magnitude and PD repetition at lower phase angles, whereas the negative skewness indicates the PD with higher magnitudes and higher repetition rate at the higher phase angles. It seems that the PD distribution with negative skewness occurs at higher AC

http://dx.doi.org/10.5772/67100

voltage, whereas the PD distribution with positive skewness occurs at lower voltage. Thus, the PD distribution with positive skewness gives more severe effect due to greater space charge activities on the wall of the electrical treeing. In addition, positive kurtosis (Ku > 3) describes that the PD has sharpened the distribution, whereas the Ku < 3 indicates that the PD distribution was flattened. However, the kurtoses of positive and negative PDs were found to be decreased over time due to enhancement of PD repetition activities. These variations of skewness and kurtosis have indicated that the distribution is skewed.

Figure 2 shows the pulse count distribution, $H_n(\varphi)$, with statistical moments. It shows that the mean phase of occurrences of positive PDs have occurred between 34° and 41°, whereas the mean phase occurrences of negative PDs have occurred between 215° and 220°. However, it is difficult to determine the burst behavior from the phase shift since the shifts are quite low in values and close to each other. Meanwhile, the other three statistical moments varied in function of time.

Similar variations of the statistical moments were found in the sample of neat silicone rubber at 40°C. Mean and standard deviation of positive and negative PD magnitudes increased over time. This indicated that the positive and negative PD magnitudes have increased with the increase of tree length. It was noticed that the values of positive and negative PD magnitudes were doubled with quite similar standard deviation compared with previous sample. The skewness and kurtosis were found to fluctuate over the times that indicated the skewed properties. The results can be graphically viewed in **Figure 3**. The abrupt changes of PD magnitudes that are called as bursts were noticed during the fluctuations of PDs distribution. This implied that the space charge activity became greater during bursts in the PD events [20]. However, the PD times were longer in this condition compared with the previous condition of temperature, 20°C.

In **Figure 4**, the statistical moments of pulse count distributions, $H_n(\varphi)$, were considered. The mean of positive PD phase occurrence occurred from 36° to 60° within the first quadrant of positive half-cycle of AC applied voltage, whereas the mean phase of occurrence of negative PDs occurred within the range of 218°–244°, within the third quadrant of negative half-cycle of AC applied voltage. Moreover, the standard deviation, skewness, and kurtosis of the positive and negative PDs showed fluctuations and variations over the times.

Meanwhile, **Figure 5** presents the statistical moments of PDs data obtained during the tree growth in neat silicone rubber at 60°C. The mean and standard deviation magnitudes of positive and negative PDs increased significantly, whereas the positive and negative PD magnitudes seem to be symmetrical. It was noticed that there are bursts of PD that indicate the rapid tree growth with reproducible behavior. The skewness and kurtosis were regularly fluctuating over time of tree growth. For comparison between PD characteristics in neat silicone rubber at 40°C and 60°C, the mean PD magnitude with its corresponding standard deviation of neat silicone rubber at 60°C were greater than PD magnitude of neat silicone rubber at 40°C. This can be elucidated that more charges are trapped at post-cure temperature of 60°C due to increase in number of crosslink networks. The other two parameters (skewness and kurtosis) seem to vary positively for negative PDs, whereas the skewness and kurtosis for positive PDs behave more negatively throughout the experimental time. These behaviors would give different pulse height distribution location and shape characteristics.

$H_n(\varphi)$ distributions presented in **Figure 6** denote that the average phase of occurrence of positive PDs had occurred within the values of 44°–93°, which indicated the first and second quadrants of positive cycle of AC applied voltage. Meanwhile, the average phase of occurrence of negative PD occurred at the third quadrant of negative cycle of AC applied voltage with the range of 223°–247°. The average phases of PDs occurrence fluctuated between the two levels. The first level is related to the periods between the bursts of the tree growth, whereas the second level is indicative of where the discharges occur during the bursts period when the average phase shifts to the lower value (14). These levels are noticed in **Figure 6**.

The average discharge magnitude, the average number of PDs per second, and the average phases of occurrence of positive and negative PDs were chosen as indicators of the PD activity. The average partial discharge magnitude was calculated as the average of the absolute value of all individual PD magnitudes, both positive and negative, recorded over the entire tree growth. The average number of PDs per second was calculated as the arithmetic means of all individual 1s data acquisition intervals during the entire tree growth phenomenon. Also, the average phases of occurrence of positive and negative PDs were calculated from the entire PD activities. Additionally, power dissipated in the sample due to PD was also considered. These five physical parameters of PDs during tree growth in neat silicone rubber at temperature of 20°C, 40°C, and 60°C are depicted in **Figures 7–9**, respectively.

Based on **Figure 7**, the mean value of positive and negative PD phases was calculated and found to be equal to 37° and 218°, respectively. Meanwhile, the average total PD numbers that underlie within the values of 687–1244 with the mean value 952. The total PD magnitudes were found to be in the range of 83–218 pC with the mean value equal to 160 pC. For the dissipated power in the sample, the values were between 0.873 and 1.40 mW with the mean value equal to 1.24 mW. It is quite difficult to identify the phase shifts from the given figure since the phases occurred very close to each other but based on the discharge magnitudes, it can be seen that the PD magnitudes have increased linearly, which denote the increased growth of electrical tree length. However, the PD numbers have reduced over time since the bigger number of PDs has occurred at the initiation stage of electrical tree growth.

Figure 8 presents the graphical description of the five statistical parameters of PD activity in neat silicone rubber at a temperature of 40°C. The average phase of positive PD value was found to be equal to 44°, whereas average negative PD phase was equal to 226°. The average total PD numbers have the values between 117 and 7571, thereby the mean value is equal to 199. For average PDs, the discharge magnitudes have the values varied from 57 to 387 pC, which results in the mean value of 280 pC. Moreover, the abrupt phase shifts occurred in the beginning of tree growth with a sharp decrease in both positive and negative PD phases. Interestingly, a similar situation was observed in PD number and dissipated power plots, which have the sudden reduction of discharge number and dissipated power. Meanwhile, the discharge magnitudes increased over times, which thereby indicated the growth of electrical tree toward the ground electrode.

For an electrical tree grown in neat silicone rubber at 60°C, the corresponding plots are depicted in **Figure 9**, where it shows that the average phases of positive PDs and negative PDs are equal to 55° and 239°, respectively. Gradual phase shifts were observed for this sample. In the case of PD number, the average total PD numbers were between 65 and 649 with the mean value 197. In addition, the average PD magnitudes were obtained within the range of 70–610 pC, and the mean value was equal to 332 pC. Meanwhile, the dissipated powers fluctuated from 51.42 to 997.4 μW with resulting mean value of 668 μW. The reproducible abrupt phase shifts were noticed in the dissipated at silicone rubber at temperature of 2 power plot and also in the corresponding plot of the PD number that indicated the significant increase in the number of PD during the bursts. These bursts can be related to the rapid increase in the radial extension of the tree at the inception stage as discussed before. It was noticed as well that the increased magnitudes of discharges became bigger as tree channels grew in length toward the ground electrode.

2.2. Analysis of PD for silicone rubber-based nanocomposite (1 wt%) at temperatures of 20°C, 40°C, and 60°C

The statistical analysis was continued for the case of silicone rubber-based nanocomposite with an addition of 1 wt% nanoclay nanofillers at temperatures of 20°C, 40°C, and 60°C. The environmental relative humidity during the test was measured and was found to have the value of 45 ± 10%.

The statistical moments are computed and shown graphically in **Figure 10**. The mean positive and negative PD amplitude distributions, $H_n(q)$ fluctuated from 0 to more than 500 pC with corresponding increase in standard deviation that deviated from 0 to 400 pC over time. Some regular bursts of PD activity were noticed in which the PD magnitudes increased abruptly. The skewness and kurtosis have regular fluctuations with the skewness having negative value

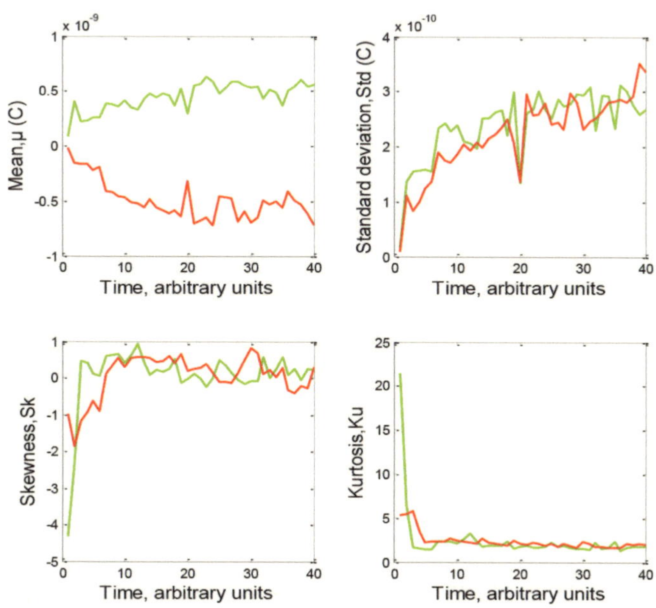

Figure 10. Mean, standard deviation, skewness, and kurtosis of positive and negative PD magnitudes, $H_n(q)$, as a function of time obtained from PDs during tree growth in silicone rubber-based nanocomposite (1 wt%) at 20°C. Green solid line = positive PD and red solid line = negative PD.

(Sk < 0), which indicated the left-skewed properties, while, the kurtosis that had values more than 3 (Ku > 3) indicated that the PD magnitude distributions have stronger peak.

The pulse count distributions, $H_n(\varphi)$, are depicted in **Figure 11**. It shows that the mean phases of occurrence of positive PDs occurred between 33° and 86°, which proves that the positive PDs occurred in the first quadrant of positive half-cycle of AC voltage. Meanwhile, the mean phases of occurrence of negative PDs had the values between 209° and 275° thereby denoting that the negative PDs occurred during the third and the forth quadrant of negative half-cycle of AC voltage. Some regular phase shifts were observed from the figure, and it was also noticed that these statistical moments increased during the beginning of PDs, and after a certain period, their values had decreased over time until the last PD.

In case of PDs during tree growth in silicone rubber-based nanocomposite with 1 wt% nanoclay nanofiller under temperature of 40°C, the mean positive and negative PD magnitude distributions, $H_n(q)$, standard deviation, skewness, and kurtosis are shown in **Figure 12**. The mean PD magnitude distributions fluctuated regularly with some repeated bursts that implied the rapid propagation of electrical tree. The skewness varied between negative and positive values, which showed left and right skewness and kurtosis were found to have values <3. Thus, the PDs magnitude distribution flattened.

The pulse count distribution, $H_n(\varphi)$ is depicted in **Figure 13**. For mean phase of occurrence of positive PDs, the minimum and maximum values are 36° and 61°, respectively. This implies that the PDs occurred at the first quadrant of positive half-cycle of AC voltage. Meanwhile, the mean

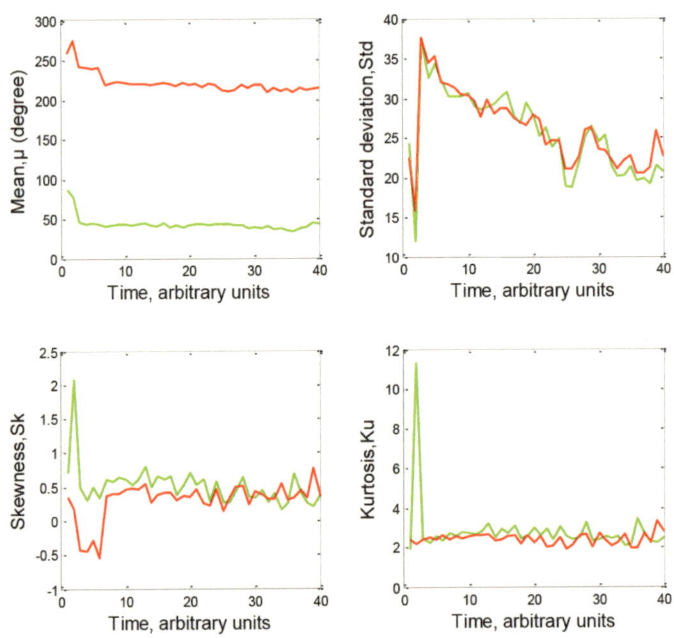

Figure 11. Mean, standard deviation, skewness, and kurtosis of positive and negative PD phase distributions, $H_n(\varphi)$, as a function of time obtained from PDs during tree growth in silicone rubber-based nanocomposite (1 wt%) at 20°C. Green solid line = positive PD and red solid line = negative PD.

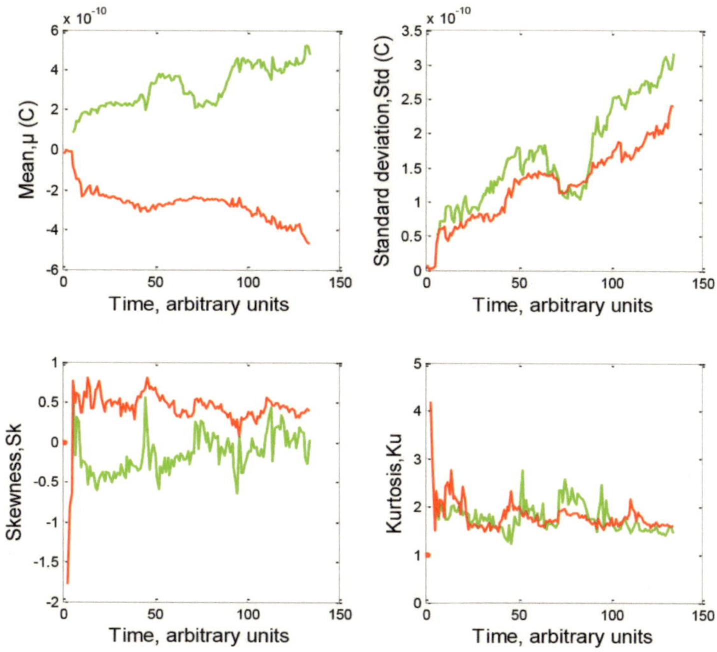

Figure 12. Mean, standard deviation, skewness, and kurtosis of positive and negative PD magnitudes, $H_n(q)$, as a function of time obtained from PDs during tree growth in silicone rubber-based nanocomposite (1 wt%) at 40°C. Green solid line = positive PD and red solid line = negative PD.

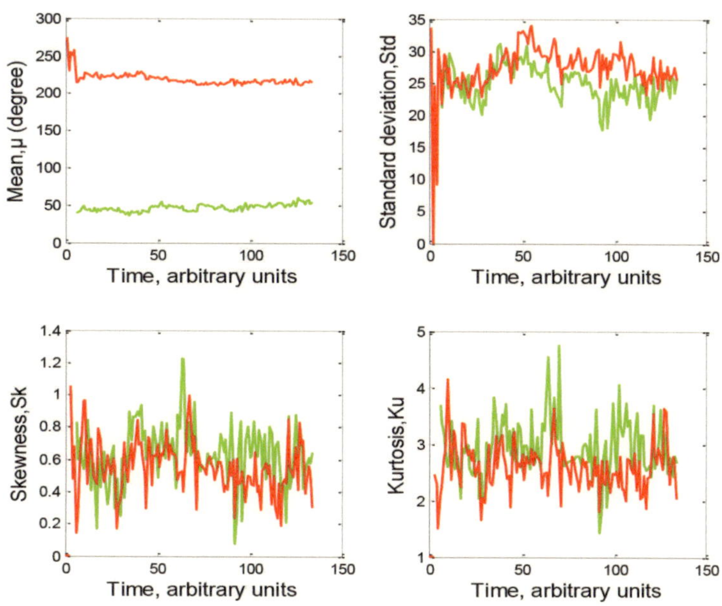

Figure 13. Mean, standard deviation, skewness, and kurtosis of positive and negative PD phase distributions, $H_n(\varphi)$, as a function of time obtained from PDs during tree growth in silicone rubber-based nanocomposite (1 wt%) at 40°C. Green solid line = positive PD and red solid line = negative PD.

phase of occurrence for negative PDs having a range of values underlying within 209°–274° implied that these PDs occurred at the third and fourth quadrant of negative half-cycle of AC applied voltage. An apparent phase shift was observed in the beginning of PD activity as a result of the rapid growth during initiation of an electrical tree. Moreover, the standard deviation, skewness, and kurtosis had varied over time in which the phases of occurrences of positive PDs have average ±25° standard deviation, whereas negative PDs possess ±27° standard deviation.

The study continued in the material at 60°C temperature to investigate the PD events during an electrical tree growth in silicone rubber-based nanocomposite. As shown in **Figure 14**, the mean and standard deviation of positive and negative PD magnitude distributions, $H_n(q)$, had increased significantly with gradual bursts occurring rapidly within short time duration. During the burst, PDs occurred throughout the whole tree structure [20]. In terms of skewness, the negative PD magnitude distribution was skewed to the right, whereas the positive PD magnitude distribution had varied between left-skewed and right-skewed. For kurtosis, most of the kurtosis values were less than three, which indicate flattened distribution.

The pulse count distribution, $H_n(\varphi)$, of PD activities in silicone rubber-based nanocomposite (1 wt%) tested under condition of higher temperature (60°C) is shown in **Figure 15**. The mean phase of occurrence of positive PDs regularly changed from 34° to 44°, whereas mean phase of occurrence of the negative PD occurred within range of 212°–225°. It was observed that the mean phases of both positive and negative PDs have small ranges which imply accumulation of PD occurrences at closer phases. However, the other three statistical moments had fluctuated regularly over times.

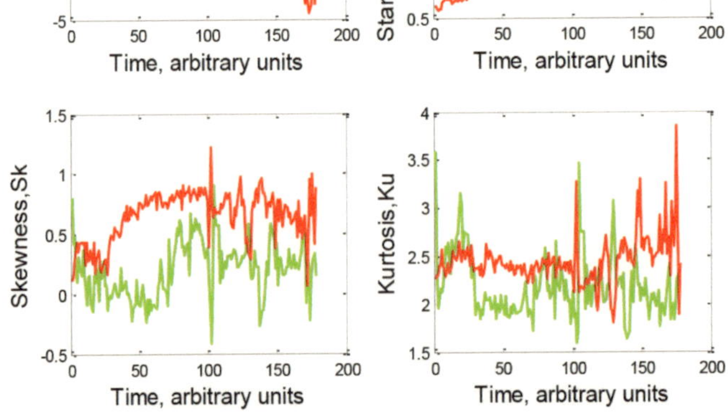

Figure 14. Mean, standard deviation, skewness, and kurtosis of positive and negative PD magnitudes, $H_n(q)$, as a function of time obtained from PDs during tree growth in silicone rubber-based nanocomposite (1 wt%) at 60°C. Green solid line = positive PD and red solid line = negative PD.

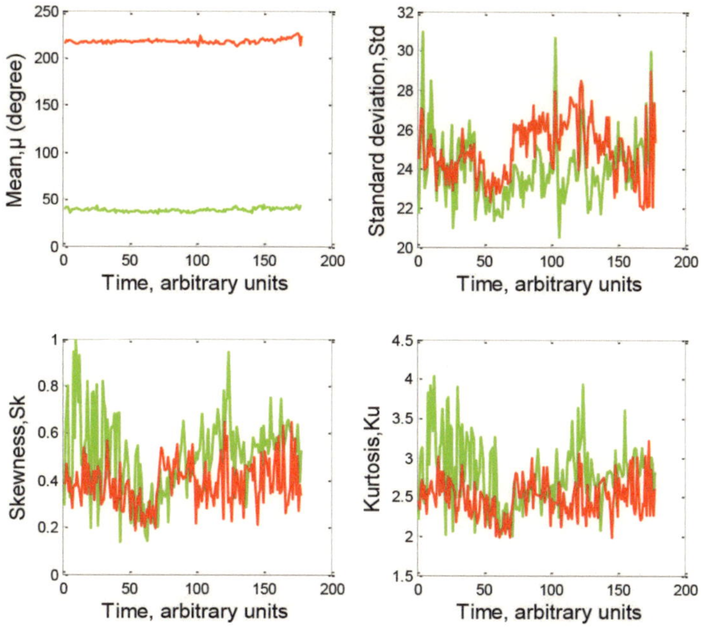

Figure 15. Mean, standard deviation, skewness, and kurtosis of positive and negative PD phase distributions, $H_n(\varphi)$, as a function of time obtained from PDs during tree growth in silicone rubber-based nanocomposite (1 wt%) at 60°C. Green solid line = positive PD and red solid line = negative PD.

Furthermore, the additional five statistical parameters of PD characteristics were studied for silicone rubber-based nanocomposite samples under three different temperatures (20°C, 40°C, and 60°C) as depicted in **Figures 16–18**, respectively. For silicone rubber-based nanocomposite tested under room temperature of 20°C as depicted in **Figure 16**, the mean value of phases of positive PDs was found to be equal to 44°, whereas for negative PDs at 223°, temporary and abrupt changes of phase shifts were observed to occur in the beginning of tree propagation with sharp increase in the PD numbers and dissipated power thereby indicating rapid growth of electrical tree as mentioned by Champion and Dodd [19]. It can be seen that after a temporary increase of the number of discharges, the PDs reduced over time with the total PD numbers having values between 249 and 1831 while the mean value was 495. On the other hand, the total PD magnitudes are found to change gradually within 31 and 641 pC. The mean total PD magnitudes were found to be equal to 450 pC. For the power dissipated in the sample, the minimum and maximum values were equal to 0.218 and 3.50 mW, respectively, with the mean value of 1.905 mW.

Furthermore, the results of PD activities in silicone rubber-based nanocomposite under temperature of 40°C are depicted in **Figure 17**, where it is shown that the average phase of positive PDs occurred at 47°. This implies that it occurred at the first quadrant of positive half-cycle of AC voltage, while the phase of occurrence of negative PDs occurred at 217°. The total number of discharges for both positive and negative PDs has a range of 1–515 with a mean value of 338. The total PD magnitudes had values that range from 5 to 492 pC with mean value of 294 pC. The minimum and maximum power dissipated in the sample equal to 0.061 and 1.5 mW,

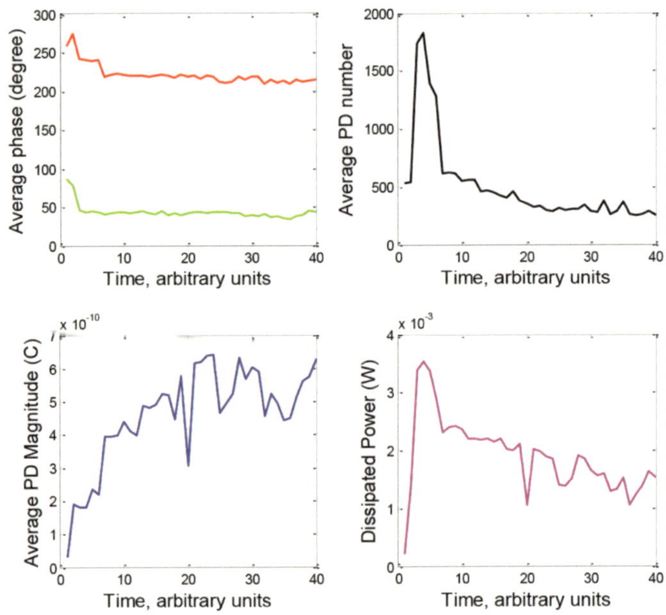

Figure 16. Average phases of occurrence of positive and negative PDs, average number of PDs, average PDs magnitude, and dissipated power as a function of time obtained from PDs activity during tree growth in silicone rubber-based nanocomposite (1 wt%) at 20°C. Green solid line = positive PD and red solid line = negative PD.

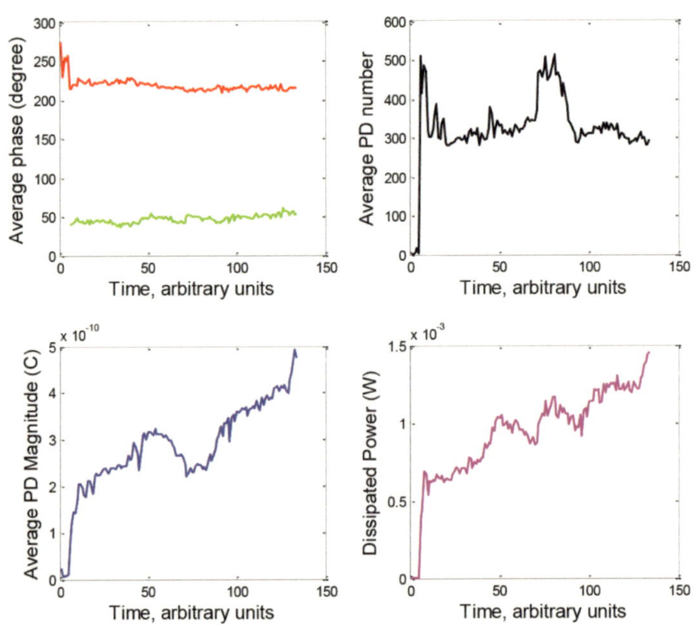

Figure 17. Average phases of occurrence of positive and negative PDs, average number of PDs, average PDs magnitude, and dissipated power as a function of time obtained from PDs activity during tree growth in silicone rubber-based nanocomposite (1 wt%) at 40°C. Green solid line = positive PD and red solid line = negative PD.

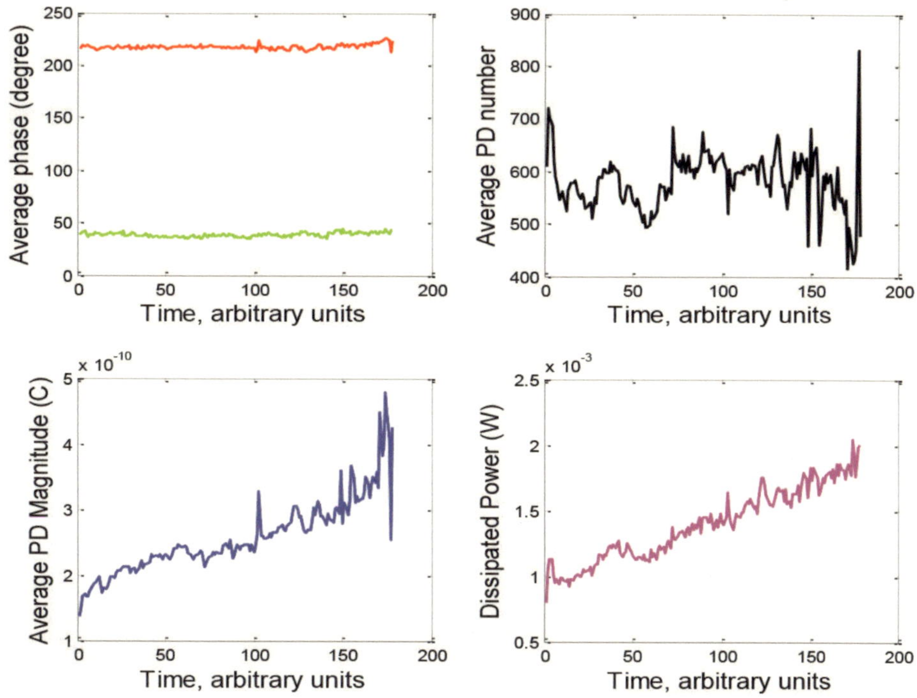

Figure 18. Average phases of occurrence of positive and negative PDs, average number of PDs, average PDs magnitude, and dissipated power as a function of time obtained from PDs activity during tree growth in silicone rubber-based nanocomposite (1 wt%) at 60°C. Green solid line = positive PD and red solid line = negative PD.

respectively, while the mean value equals to 0.936 mW. In addition, the sudden phase shift was observed to occur at the beginning of electrical tree propagation during the burst periods. Moreover, bursts situation can be observed to have occurred in the all four graphs. The abrupt changes in the total PD numbers imply that the growth of an electrical tree was faster during inception stage that had enhanced the PD repetition events.

In the case of the sample at 60°C temperature, the PD characteristics are illustrated in **Figure 18**. The mean values of the phases of positive PD and negative PD were obtained and are equal to 38° and 217°, respectively, which imply the positive and negative PDs occurred at first and third quadrant of the AC voltage cycles. In addition, the total positive and negative PD numbers were found to have range of values between 416 and 831 with mean value equal to 580. In terms of PD magnitudes, total PDs had fluctuated in the ranges of 137–478 pC with the mean value of 257 pC. It was also found that the power dissipated in the sample has minimum, maximum, and mean values of 0.795, 2.0, and 1.395 mW, respectively. The phase shift was noticed to occur during the short time of burst. Moreover, plenty of burst-type behaviors were identified, and it can be seen that the plot of PD numbers has increased significantly during the bursts. The bursts can be seen in the plot of PD magnitudes and dissipated power as well, and the discharge magnitudes and the power increased with the time of tree propagation.

2.3. Analysis of PD for silicone rubber-based nanocomposite (3 wt%) at temperatures of 20°C, 40°C, and 60°C

In the case of silicone rubber-based nanocomposite with an addition of 3 wt% of nanoclay nanofiller, it was found that the PDs have occurred and were repeated at shorter time in arbitrary units of time than the actual time of voltage application (4 hours). Similarly, the statistical moments of PD magnitude distribution, $H_n(q)$ and PD phase distributions, $H_n(\varphi)$ in silicone rubber-based nanocomposite (3 wt%) at 20°C, 40°C, and 60°C temperature are illustrated in **Figures 19(a)–(c), 20(a)–20(c),** respectively. At 20°C, the positive and negative PD magnitude distributions were found to fluctuate over time. The mean and standard deviation exhibited temporary changes of amplitudes. In terms of skewness and kurtosis, both of them fluctuated, which indicated that the distribution have skewed properties and sharper than normal distribution, respectively. In case of $H_n(\varphi)$, the mean, standard deviation, skewness, and kurtosis have varied values as functions of time thereby showing small amplitude of PDs and some dashed lines indicating intermittent/sporadic PDs.

The statistical moments of pulse magnitude distributions and pulse count distributions for the test performed at a temperature of 40°C are shown in **Figure 19(b)** and **Figure 20(b),** respectively. All the statistical moments fluctuated, and some of them have dashed lines, which show sporadic PDs. The mean of positive and negative PDs have fluctuated between the absolute values of 2 and 6 pC. Thus, the amplitudes of the mean and standard deviation were considered very low compared with the previous cases discussed in the Sections 2.1 and 2.2. The skewness and kurtosis show variations over the time. It is considered that the positive PD distribution was right-skewed and the negative PD was left-skewed, while the positive and negative PD distributions are sharper than normal distributions. The mean phases of occurrence of positive and negative PDs occurred in the whole quadrants of AC voltage cycle.

Under the temperature of 60°C, both PD magnitude and count distributions have sporadic behaviors as can be seen in **Figures 19(c)** and **20(c)**, where it shows that both negative and positive PDs only occurred at different times. It can be said that during both the half-cycle of AC voltage either positive or negative PDs occur independently at certain duration of times. The skewness and kurtosis distributions were found to have dashed lines as well. It can be said that the positive PD distributions were right-skewed and the negative PD distributions were left-skewed. Meanwhile, the kurtosis of positive and negative PD distributions was more prone to be flatter than normal distribution.

In terms of physical parameters of PD occurrences, the results are depicted in **Figures 21(a)–(c)** for the case of 20°C, 40°C, and 60°C, respectively. It was found that in the average phase of PD occurrence of all the cases (20°C, 40°C, and 60°C) were found to be relatively insensitive. The PD number and PD magnitude have shown very low values in terms of amplitudes compared with the PD results of previous two sections. The sporadic PDs affected the results. It implied that the PD occurrences were material-dependent since there were no electrical trees observed but only weak light emission, which was considered as electroluminescence. The applied AC voltage of 10 kV_{rms} was insufficient to create an electrical tree. However, PD was detected to have occurred sporadically in very low amplitudes of number and magnitude of PDs. The reasons for all these PD characteristics are discussed thoroughly in the following sections.

http://dx.doi.org/10.5772/67100

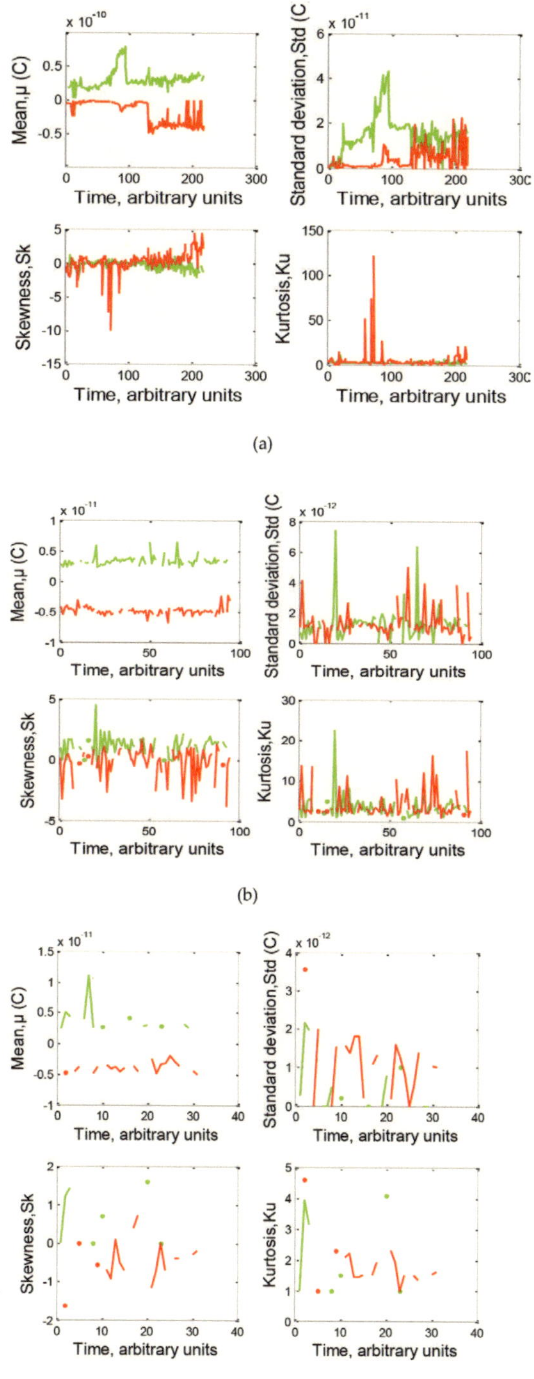

Figure 19. Mean, standard deviation, skewness, and kurtosis of positive and negative PD magnitudes, $H_n(q)$, as a function of time obtained from PDs during tree growth in silicone rubber-based nanocomposite (3 wt%) at (a) 20°, (b) 40°, and (c) 60°C. Green solid line = positive PD and red solid line = negative PD. Dashed lines indicate sporadic PDs.

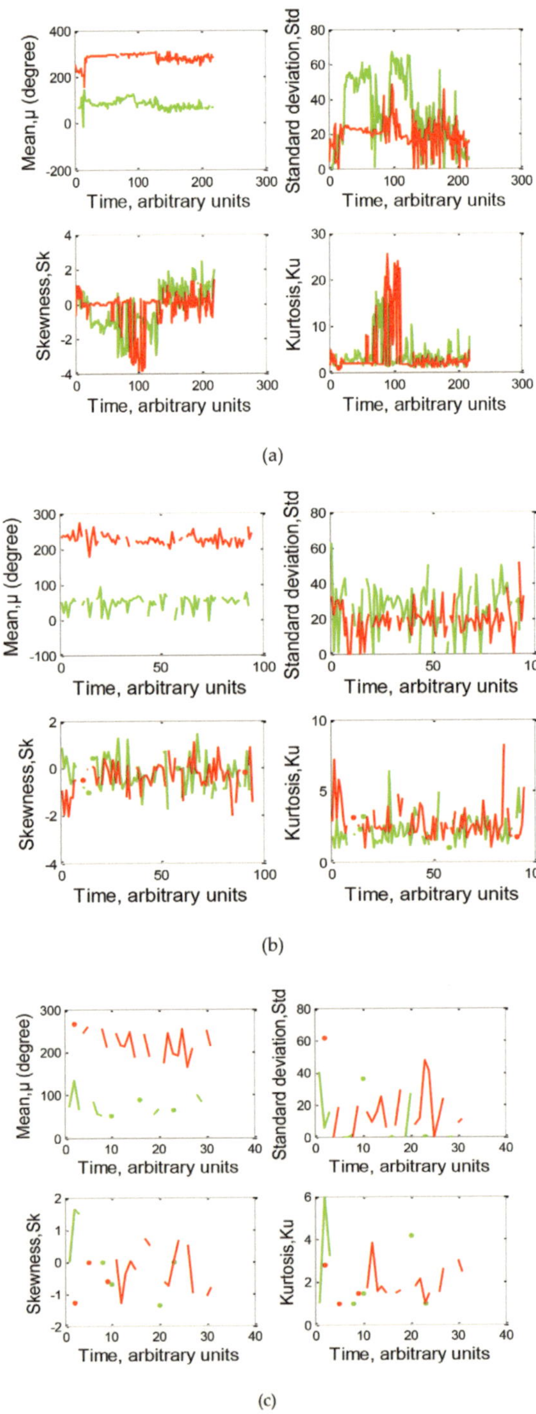

Figure 20. Mean, standard deviation, skewness, and kurtosis of positive and negative PD phase distributions, $H_n(\phi)$, as a function of time obtained from PDs during tree growth in silicone rubber-based nanocomposite (3 wt%) at (a) 20°C, (b) 40°C, and (c) 60°C. Green solid line = positive PD and red solid line = negative PD. Dashed lines indicate sporadic PDs.

http://dx.doi.org/10.5772/67100

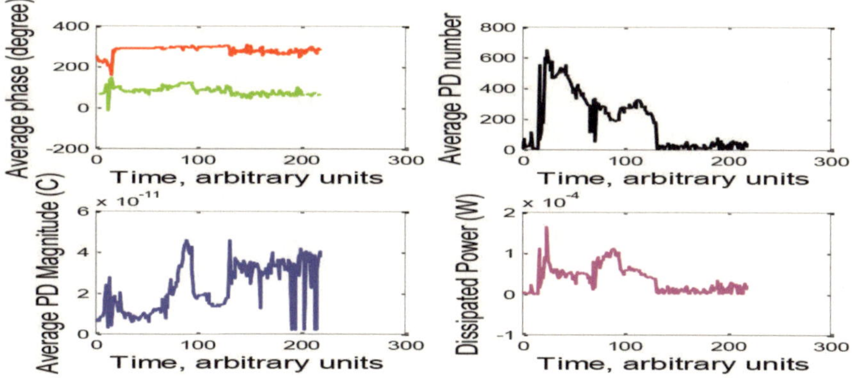

(a)

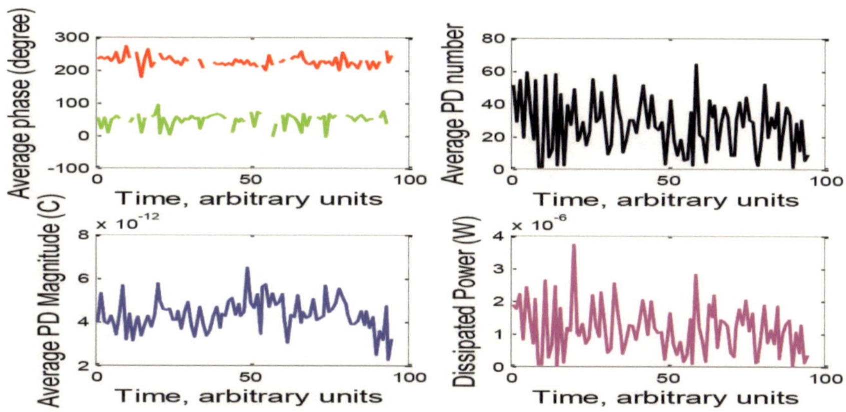

(b)

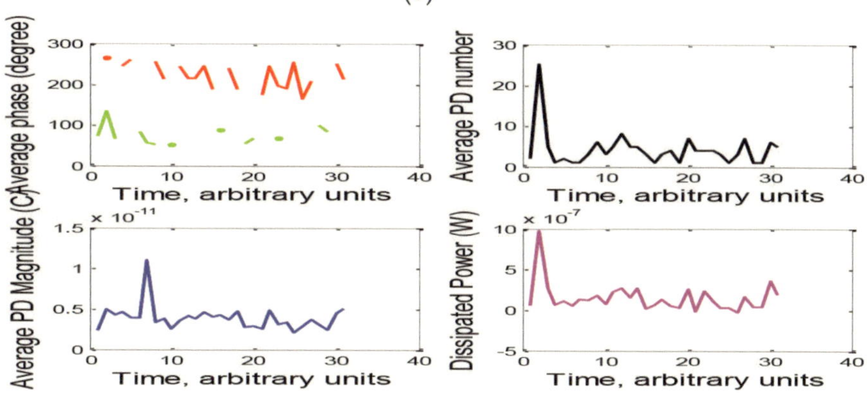

(c)

Figure 21. Average phases of occurrence of positive and negative PDs, average number of PDs, average PDs magnitude, and dissipated power as a function of time obtained from PDs activity during tree growth in silicone rubber based nanocomposite (3 wt%) at (a) 20°C, (b) 40°C, and (c) 60°C. Green solid line = positive PD and red solid line = negative PD. Dashed lines indicate sporadic PDs.

3. Discussion

As discussed by Champion et al. [20] and Champion and Dodd [19], PD events in neat epoxy resin (CY1311) exhibited burst-type behavior. During the burst period, the number of discharges increases significantly, which indicate faster growth of electrical tree, and the discharges were found to be reproducible due to electrical tree length elongation. In addition, it was found that the phase of occurrences shifted to lower values during the burst periods, suggesting the enhancement and fluctuation of local electric field during the half-cycle that leads to the occurrence of PD at lower angle [21].

The processing results of the PD data are shown in **Table 1**. This presents a summary of the recorded PD data. In the case of neat SiR, the average PD magnitude was found to increase with increasing temperature, while the average number of PDs per second was found to decrease with increasing temperature. Also, a significant shift in the average phase of PD occurrence of both the positive and negative discharge distributions was taking place simultaneously. The average phase of occurrence of the positive PDs increased from 37° at 20°C to 55° at 60°C. A similar dependence on temperature of the PD phases was found in the case of the electrical trees growth in a flexible epoxy resin. This phase shift is attributed to an increase in the electrical conductivity of the epoxy resin with increasing temperature leading to decreased accumulation of space charge surrounding the tree structure.

Type of sample	PD characteristics	20°C			40°C			60°C		
		Min	Max	Ave	Min	Max	Ave	Min	Max	Ave
Neat SiR	Discharge magnitude (pC)	83	218	160	57	387	280	70	610	332
	Number of PDs (s^{-1})	687	1244	952	117	7571	199	65	649	197
	Phase of occurrence of positive PDs (°)	34	41	37	36	60	44	44	93	55
	Phase of occurrence of negative PDs (°)	215	220	218	218	244	226	223	247	239
	Dissipated power (mW)	0.87	1.40	1.24	0.35	3.4	0.54	0.05	0.997	0.668
SiR/1 wt% nanoclay	Discharge magnitude (pC)	31	641	450	5	492	294	137	478	257
	Number of PDs (s^{-1})	249	1831	495	1	581	338	416	831	580
	Phase of occurrence of positive PDs (°)	33	86	44	36	61	47	34	44	38
	Phase of occurrence of negative PDs (°)	209	275	223	209	274	217	212	225	217
	Dissipated power (mW)	0.22	3.50	1.91	0.061	1.5	0.94	0.75	2.00	1.395
SiR/3 wt% nanoclay	Discharge magnitude (pC)	2	45	22	2	6	4	2	11	4
	Number of PDs (s^{-1})	1	643	181	1	64	27	1	25	4
	Phase of occurrence of positive PDs (°)	17	142	80	5	94	46	52	133	75
	Phase of occurrence of negative PDs (°)	154	306	280	177	272	226	164	265	222
	Dissipated power (μW)	0.34	163	35.7	0.002	3.7	1.13	0.03	0.98	0.15

Min is minimum value, Max is maximum value, and Ave is average value.

Table 1. Summary of partial discharge statistical parameters.

However, the reverse circumstances on PD numbers and PD magnitudes have been exhibited in epoxy resin sample since the PD numbers increased while PD magnitudes decreased with the increasing of temperature from 20°C to 70°C [21]. This dissimilarity between epoxy resin and silicone rubber is probably due to the increase of vulcanization/cross-link numbers in silicone rubber matrices with increase of temperature which was discussed thoroughly by Du et al. [22].

In the case of 1 wt% nanoclay nanocomposite, the average PD magnitude has been found to decrease with increasing temperature from 450 pC at 20°C to 257 pC at 60°C. However, the other three PD physical parameters remained relatively insensitive to the changes in temperature. The decrease in PD magnitude when the temperature increased in silicone rubber-based nanocomposites (1 wt%) is because the addition of nanoparticles contributes to trapping of space charges in which the electrons injected from the electrode move toward the opposite electrode and get trapped at the nanoparticle surfaces and inside the nanoparticles. When the applied field is sufficiently high, it can cause ionization of nanoparticle within the polymer. Heterocharges will occur due to discharge (bulk material transport charges more easily than they are extracted by the polymer/metal interface). Thus, internal field will be created to oppose the direction of applied field, making the net field inside the nanoparticle more than the discharge inception field thereby extinguishing the discharge. This would reduce the PD magnitude. If the applied field is high enough and exceeds the internal field, the discharge will occur again. However, the particles tend to agglomerate thereby resulting in less interfacial area leading to lesser charge carriers getting trapped around the particle surfaces (localized states). Due to repetition and sufficiently high electric field, the trap sites provided by microparticles for electron localizations are not enough thereby leading to drift of electron (detrapping). This would cause charge movements (mobility) that give more PD repetition rate.

The addition of 3 wt% nanoclay nanoparticles in silicone rubber had caused an effective electrical tree inhibition. Electrical trees have not been observed during the 4 hours of the experiment. Only localized light emission was detected at the needle tip using the CCD camera, which is recognized as electroluminescence. It appears that the silicone rubber matrices were reinforced by an addition of 3 wt% of nanoclay nanofillers to an extent that prevented the formation of an initial void with the sufficient dimension required for PD generation and subsequent electrical tree initiation. The PD magnitudes and PD numbers were obtained, and they decrease with the increase in temperature from 20°C to 60°C: the PD magnitudes decreased from 22 to 4 pC. However, the phase of positive PDs was found to be insensitive to temperature changes, whereas the phase of negative PDs shifted to the lower values with the increase in temperature.

Moreover, the small discharges detected in that nanocomposite material was probably a swarming partial micro-discharges (SPMD) which is related to the existence of microvoids at polymer/electrode interface due to imperfect bonding between the tungsten needle and the silicone rubber. Tanaka [23] reported that small PD of less than 1 pC was detected in the void prior to tree initiation. Swarming partial micro-discharge phenomena can be found in [23–26]. Thus, the detection of PD prior to the tree initiation may be caused by the microvoid at the needle tip due to charge carriers recombination. The holes and electrons are injected into the polymer during the positive and negative half-cycle of AC voltage, and they are trapped in the recombination centers. Light is emitted when the electron-hole recombination occurs at the luminescent center [27]. However, the electric field intensity was insufficient to initiate the electrical treeing due to low energy. In addition, it appeared that the silicone rubber matrices were reinforced by the addition of 3 wt% of nanoclay nanofiller to an extent that prevented the formation of a void with sufficient dimensions required for PD generation and electrical tree initiation.

In polymer nanocomposites, the nanoparticles act as charge carrier traps with high barrier potentials. These trapped carriers need more energy to get extracted from one trap to another trap; therefore, this would slow down the growth of electrical treeing. With the addition of 1 wt% of nanoclay nanofillers, the tree inception time and propagation time were enhanced distinctly. However, PD magnitudes and PD numbers of silicone rubber-based nanocomposite (1 wt%) became insensitive with the increase in temperature as the PD magnitudes seem to be higher than neat silicone rubber at 20°C and 40°C, whereas the PD numbers of silicone rubber-based nanocomposite (1 wt%) were greater than neat silicone rubber at 40°C and 60°C. The reason of the increase in PD magnitudes and PD numbers is attributable to agglomeration of nanofiller in silicone rubber matrix. The tactoids represent the weak points that could enhance the PD repetition inside the polymer. However, the agglomerated nanoparticles bigger than 100 nm are considered as a microparticle. The microparticle has also contributed to the suppression of an electrical tree as reported in reference [17, 28] that reacted as a physical barrier to the growth of tree channel resulting in the tree forming more side branches when it collides with the microparticle.

Author details

Mohd Hafizi Ahmad

Address all correspondence to: mohdhafizi@utm.my

Institute of High Voltage and High Current, Faculty of Electrical Engineering, Universiti Teknologi Malaysia, Johor Bahru, Johor, Malaysia

References

[1] Andritsch T, Kochetov R, Lennon B, Morshuis PHF, Smit JJ. Space charge behavior of magnesium oxide filled epoxy nanocomposites at different temperatures and electric field strengths. In: Proceedings of the IEEE Electrical Insulation Conference (EIC); 2011. pp. 136–140.

[2] Bamji S, Bulinski A, Abou-Dakka M, McIntyre D. Space charge in polypropylene containing synthetic nanoparticles. In: IEEE Annual Report of the Conference on Electrical Insulation and Dielectric Phenomena; 2009.pp.662–665.

[3] Alapati S, Thomas MJ: Electrical treeing and the associated PD characteristics in LDPE nanocomposites. IEEE Trans Dielectr Electr Insul. 2012;**19**(2):697–704. DOI: 10.1109/TDEI.2012.6180265.

[4] Preetha P, Thomas M: Partial discharge resistant characteristics of epoxy nanocomposites. IEEE Trans Dielectr Electr Insul. 2011;**18**(1):264–274. DOI: 10.1109/TDEI.2011.5704518.

[5] Guastavino F, Dardano A, Squarcia S, Tiemblo P, Guzman J, Benito E. Electrical treeing in LDPE nanocomposite materials. In: Proceedings of the IEEE Conference on Electrical Insulation and Dielectric Phenomena; 2009. pp. 697–700.

[6] Sridhar A, Joy Thomas M. Electrical treeing in polyethylene: effect of nanofillers on tree inception and growth. In: Proceedings of the IEEE International Conference on High Voltage Engineering and Application; 2010. pp. 576–579.

[7] Tanaka T, Kozako M, Fuse N, Ohki Y: Proposal of a multi-core model for polymer nanocomposite dielectrics. IEEE Trans Dielectr Electr Insul. 2005 Aug;**12**(4):669–681. DOI: 10.1109/TDEI.2005.1511092.

[8] Du BX, Gao ZLM, Han T. Effect of temperature on electrical tree in silicone rubber. In: Proceedings of the 10th IEEE International Conference on Solid Dielectrics; 2010. pp. 1–4.

[9] Shimizu N, Shibata Y, Ito K, Imai K, Nawata M. Electrical tree at high temperature in XLPE and effect of oxygen. In: Proceedings of the IEEE Annual Report Conference on Electrical Insulation and Dielectric Phenomena; 2000. pp. 329–332.

[10] Jongen R, Gulski E, Smit J. Failures of medium voltage cable joints in relation to the ambient temperature. In: Proceedings of the CIRED 20th International Conference and Exhibition on Electricity Distribution—Part 1; 2009. pp. 8–11.

[11] Chen XR, Hu LB, Xu Y, Cao XL, Gubanski SM. Investigation of temperature effect on electrical trees in XLPE cable insulation. in: IEEE Annual Report Conference on Electrical Insulation and Dielectric Phenomena; 2012. pp. 612–615.

[12] Dodd SJ, Chalashkanov N, Fothergill JC. Statistical analysis of partial discharges from electrical trees grown in a flexible epoxy resin. In: IEEE Annual Report Conference on Electrical Insulation and Dielectric Phenomena; 2008. pp. 666–669.

[13] Gulski E: Computer-aided measurement of partial discharges in HV equipment. IEEE Trans Electr Insul. 1993;**28**(6):969–983. DOI: 10.1109/14.249370.

[14] Chalashkanov N. Influence of water absorption and temperature on charge transport and electrical degradation in epoxy resins [Thesis]; Department of Engineering, University of Leicester; 2011.

[15] Sahoo NC, Salama MM, Bartnikas R: Trends in partial discharge pattern classification: a survey. IEEE Trans Dielectr Electr Insul. 2005;**12**(2):248–264. DOI: 10.1109/TDEI.2005.1430395.

[16] Kurnianto R, Murakami Y, Hozumi N, Nagao M: Characterization of tree growth in filled epoxy resin: the effect of filler and moisture contents. IEEE Trans Dielectr Electr Insul. 2007;**14**(2):427–435. DOI: 10.1109/TDEI.2007.344623.

[17] Kurnianto R, Murakami Y, Hozumi N, Nagao M. Characterization of tree growth in inorganic-filler/epoxy resin composite material. In: Proceedings of IEEE 8th International Conference on Properties and applications of Dielectric Materials; 2006. pp. 123–126.

[18] Patsch R: Electrical and water treeing: a chairman's view. Electr Insul IEEE Trans. 1992;**27**(3):532–542. DOI: 10.1109/14.142716.

[19] Champion JV, Dodd SJ: Systematic and reproducible partial discharge patterns during electrical tree growth in an epoxy resin. J Phys D Appl Phys. 1996;**29**(3):862–868. DOI: 10.1088/0022–3727/29/3/053.

[20] Champion JV, Dodd SJ, Alison JM: The correlation between the partial discharge behaviour and the spatial and temporal development of electrical trees grown in an epoxy resin. J Phys D Appl Phys. 1996;**29**(10):2689–2695. DOI: 10.1088/0022–3727/29/10/023.

[21] Chalashkanov NM, Dodd SJ, Dissado LA, Fothergill JC. Pulse sequence analysis on PD data from electrical trees in flexible epoxy resins. In: IEEE Annual Report Conference on Electrical Insulation and Dielectric Phenomena; 2011. pp. 776–779.

[22] Du B, Ma Z, Gao Y, Han T: Effect of ambient temperature on electrical treeing characteristics in silicone rubber. IEEE Trans Dielectr Electr Insul. 2011;**18**(2):401–407. DOI: 10.1109/TDEI.2011.5739443.

[23] Tanaka T: Internal partial discharge and material degradation. IEEE Trans Electr Insul. 1986;EI-**21**(6):899–905. DOI: 10.1109/TEI.1986.348999.

[24] Danikas M: Small partial discharges and their role in insulation deterioration. IEEE Trans Dielectr Electr Insul. 1997;**4**(6):863–867. DOI: 10.1109/94.654733.

[25] Danikas MG: The definitions used for partial discharge phenomena. IEEE Trans Electr Insul. 1993;**28**(6):1075–1081. DOI: 10.1109/14.249381.

[26] Tanaka T: Partial discharge pulse distribution pattern analysis. IEE Proc—Sci Meas Technol. 1995;**142**(1):46–50. DOI: 10.1049/ip-smt:19951561.

[27] Takahashi T, Suzuki J, Miyata H, Iemura S, Itoh T, Nakiri T: Relation between electroluminescence and degradation in XLPE. IEEE Trans Dielectr Electr Insul. 2001;**8**(1):91–96. DOI: 10.1109/94.910430.

[28] Nagao M, Oda K, Nishioka K, Muramoto Y, Hozumi N. Effect of filler on treeing phenomenon in epoxy resin under AC voltage. In: Proceedings of IEEE International Symposium on Electrical Insulating Materials (ISEIM '01); 2001. pp. 611–614.

Chapter 6

Effect of Stretching on Electrical Properties of Low Density Polyethylene/MgO Nanocomposites

Youyuan Wang and Can Wang

Additional information is available at the end of the chapter

http://dx.doi.org/10.5772/66982

Abstract

Pulsed electroacoustic (PEA) method, dielectric analyzer, and high resistance meter were used to research the DC breakdown, space charge behavior, permittivity, loss tangent, and volume resistivity of neat low density polyethylene (LDPE) and LDPE/MgO nanocomposites with concentration of 1, 3, and 5 wt% with and without the elongation ratio of 1.1. Results indicate that the DC breakdown strengths of neat LDPE and nanocomposites decrease after stretching. The heterocharges near electrodes in neat LDPE change to homocharges after stretching and a large amount of positive and negative charges accumulated in samples with concentration of 3 and 5 wt%. Meanwhile, homocharges near cathode electrode in nanocomposite with concentration of 1 wt% decrease a little and a small amount of positive charges were observed in the samples after stretching. Furthermore, there are different increases in amplitudes of permittivity in all the samples after stretching, as well as the loss of tangents especially in the frequency domain from 10^{-1} to 10^{2} Hz. The results of volume resistivity show that comparing with the nanocomposite with the concentration of 1 wt% whose volume resistivity decrease after stretching, larger volume resistivity is observed in neat LDPE and nanocomposites with the concentration of 3 and 5 wt%.

Keywords: stretching, LDPE/MgO nanocomposites, interface, free volume, electrical properties

1. Introduction

Polyethylene was widely used in high voltage equipment especially in DC cables because of its excellent electrical and mechanical properties [1, 2]. In comparison with the polyethylene, polymer-based nanocomposites have better electrical and mechanical properties, such

INTECH
open science | open minds

© 2017 The Author(s). Licensee InTech. This chapter is distributed under the terms of the Creative Commons Attribution License (http://creativecommons.org/licenses/by/3.0), which permits unrestricted use, distribution, and reproduction in any medium, provided the original work is properly cited.

as higher DC or AC breakdown strength and volume resistivity, smaller amount of space charge accumulation, and so on [3–6]. However, polyethylene insulation materials may be affected by stretching during operation, which may lead to the change of electrical properties. Therefore, some questions arise such as: what are the properties of polyethylene and nanocomposites after stretching? And how does stretching affect the electrical properties of polyethylene and nanocomposites?

It is generally agreed that electrical properties of polymer insulation materials can improve after the addition of nanoparticles. In order to analyze the mechanism, nanocomposites with different nanoparticles and different polymer matrixes were prepared, such as LDPE/ZnO nanocomposites, LDPE/SiO_2 nanocomposites, Epoxy/TiO_2 nanocomposites, Epoxy/POSS nanocomposites, and so on [7–12]. According to the research, it is concluded that the interface between nanoparticles and polymer matrix can dominate the characteristics of polymer-based nanocomposites [13–18]. On this basis, some interface models such as multicore model and space charge model have been proposed [19–21].

Recent reports on the effects of stretching on electrical properties of polymer focus on carbon nanotube/polymer-based materials. To the best of our knowledge, there are currently no reports on the effect of stretching on the electrical properties of polymer-based nanocomposites. Previous studies have indicated that stretching can change the space charge behavior of poly(vinyl chloride) (PVC) and cross-linked polyethylene (XLPE), and even the conformations of PVC [22, 23]. In carbon nanotube/polymer composites, stretching causes deflection of the carbon nanotubes and the deflection extent can reach up to 45°, decreasing the conductivity of the carbon nanotube/polymer [24–27]. Additionally, influences of hydrostatic pressure on the dielectric properties of polyethylene/aluminum oxide nanocomposites have been researched, with results showing that permittivity and free volume will decrease after pressure [28].

Stretching may change the interface between nanoparticles and polymer matrix. This research can enhance our understanding of electrical properties of polymer-based nanocomposites. Furthermore, in order to answer the questions mentioned above, the DC breakdown, space charge behavior, permittivity, loss tangent, and volume resistivity of neat LDPE and LDPE/MgO nanocomposites with concentration of 1, 3, and 5 wt% with and without elongation ration of 1.1 were researched in this chapter. On this basis, the change of interface, space charge traps, and free volume were discussed.

2. Experiment

2.1. Material

The base polymer used was an additive-free low density polyethylene (LDPE) with a melt flow index of 2.1–2.2 g/min and a density of 0.910–0.925 mg/cm^3. The nanoparticles used were MgO with a 40-nm particle size. Nanocomposites, with concentration of 1, 3, and 5 wt%, were press-modeled at 393 K and at a pressure of 10 MPa, to produce films with a thickness of about 190 μm. Samples were degassed in a vacuum drier for 7 days at 353 K. Dumbbell-shaped samples cut by a dumbbell cutting tool were stretched in tension machine. The elongation ratio was chosen as 1.1–1.4 in the previous research [22, 23]. However, the elongation ratio of insulation material of cables may not be so high. Therefore, we choose 1.1 as the elongation ratio in this chapter. Additionally, because the polyethylene is an elastic material, the elongation ratio of samples in the tension machine should be higher than 1.1 to ensure that the elongation ratio of samples is about 1.1 after rebound.

2.2. Characteristics

2.2.1. Electrical breakdown strength testing

Samples were placed between spherical electrodes (ø = 6.3 cm), which were immersed in transformer oil to avoid surface flashover on the samples. The applied DC voltage was increased linearly at a rate of 0.5 kV/s until breakdown occurred. The thickness of the tested samples was measured to calculate the DC breakdown strength. Each sample was tested 15 times and the two-parameter Weibull distribution was used to analyze the DC breakdown characteristics of the samples.

2.2.2. Space charge measurements

The space charge measurements were carried out with a pulsed electroacoustic (PEA) system, whose pulse width is 2–5 ns, pulse amplitude is 200 V, and output voltage is 0–20 kV. All samples were measured at room temperature (25 ± 1°C). In this chapter, a DC electrical field was applied for 1 h and the space charge formation was confirmed with the polarization of 10 s, 30 s, 1 min, 5 min, 10 min, 15 min, 20 min, 25 min, 30 min, 35 min, 40 min, 50 min, and 1 h. The reproducibility of the space charge behavior was confirmed by repeating the experiments five times for each group.

2.2.3. Dielectric properties

The dielectric properties of the materials were measured in the frequency domain from 10^{-1} to 10^{6} Hz at room temperature (25 ± 1°C) by using a Novocontrol ALPHA-A high resolution dielectric analyzer. Prior to dielectric analysis, gold electrodes were deposited onto both surfaces of the specimens by sputtering. The diameter of sputtered electrodes is 15 mm. The results are the average measurements of five different specimens for each sample and the error in the measurement is within 2%.

2.2.4. Volume resistivity

Volume resistance was measured by a high resistance meter (model PC68). Compression-molded sheet having a diameter of 100 mm was inserted into the sample holder and charged for 1 min at 500 V. The volume resistance measurements were carried out at room temperature (25 ± 1°C). Volume resistivity (ohm. m) = 21.237 (R_v/t), where R_v is the volume resistance (ohm) and t is the thickness (m) of the sheet. The results are the average measurements of five different specimens for each sample.

3. Results

3.1. DC breakdown

The DC breakdown properties of insulation are affected by many factors. Randomness and dispersion of DC breakdown cannot be prevented even under identical conditions. It is recognized that the two-parameter Weibull probability can fit the DC breakdown strength of solid materials effectively with the following expression [29]:

$$F(U;\alpha,\beta) = 1-\exp\left(-\left(\frac{U}{\alpha}\right)^{\beta}\right) \tag{1}$$

where U represents the DC breakdown strength; α represents scale parameters, indicating the DC breakdown strength when the failure rate is 0.632; and β represents shape parameters, indicating the randomness of the data.

Figure 1 is the Weibull probability plot for the DC breakdown of neat LDPE and LDPE/MgO nanocomposites before and after stretching, and the scale parameters and shape parameters are listed in **Tables 1** and **2**.

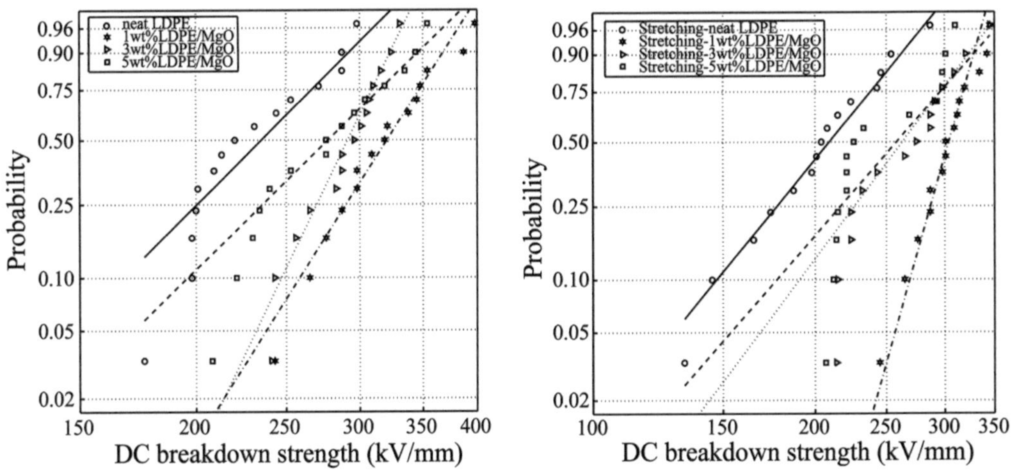

(a) (b)

Figure 1. Weibull probability plot for DC breakdown strength of LDPE and LDPE/MgO nanocomposites before and after stretching. (a) Before stretching and (b) after stretching.

	LDPE	1 wt% LDPE/MgO	3 wt% LDPE/MgO	5 wt% LDPE/MgO
α (kV/mm)	248.76	337.71	302.07	297.92
β	6.65	8.20	13.17	6.89

Table 1. Scale parameter α and shape parameter 72 for LDPE and LDPE/MgO nanocomposites before stretching.

	LDPE	1 wt% LDPE/MgO	3 wt% LDPE/MgO	5 wt% LDPE/MgO
α (kV/mm)	220.59	314.39	286.05	266.32
β	5.62	12.99	7.41	6.99

Table 2. Scale parameter α and shape parameter 72 for LDPE and LDPE/MgO nanocomposites after stretching.

According to **Figure 1**, the scale parameters α increase after adding MgO nanoparticles, especially the nanocomposites with a concentration of 1 wt%, whose scale parameter is 337.71 kV/mm. However, the scale parameters α decrease with the increase of concentration of nanoparticles. This phenomenon is the same as the research of Murakami [30]. Furthermore, the scale parameters of all samples decrease after stretching.

On the basis of the space charge behavior, the heterocharges near electrodes in neat LDPE change to the homocharges after adding nanoparticles, which means the electrical strengths between electrodes and materials increase as well as electron injection currents decrease. Therefore, the amount of electrons or holes injected by electrodes may increase and lead to the improvement of DC breakdown strengths [31]. Additionally, the interface between nanoparticles and polymer matrix may introduce a large amount of charge traps [32]. As the carriers are trapped, the energy of nanofillers may decrease and may also promote the DC breakdown strength.

3.2. Space charge behavior

Figure 2 is the space charge behavior of neat LDPE and LDPE/MgO nanocomposites. Before stretching, heterocharges near electrodes are observed in neat LDPE with the largest amount near anode, whose value is 2.7 C/m^3 after polarization time of 10 s. Meanwhile, there are positive packet-like space charges moving from anode to cathode in neat LDPE with the value of 3.5 and 3.6 C/m^3 with the polarization time of 10 and 30 s, respectively. This phenomenon is the same as the research of Murakami [30]. Adding nanoparticles, homocharges are observed near electrodes in nanocomposites, whose amount increases with the increase of polarization times, especially in nanocomposites with concentration of 1 and 5 wt%. After all, all samples prepared in this chapter can suppress space charge accumulation and in which the nanocomposite with concentration of 3 wt% has the best effect.

After stretching, heterocharges near electrodes in neat LDPE change to homocharges. Space charge behavior in nanocomposite with concentration of 1 wt% changes a little and a little more positive space charges are observed. At the same time, there are positive and negative charges in nanocomposites with concentration of 3 and 5 wt% after stretching, especially in nanocomposite with concentration of 5 wt%, whose amount of positive charges is 6 C/m^3 and the amount of negative charges is 7 C/m^3. Interestingly, the space charge behavior of nanocomposite with concentration of 3 wt% after stretching is very similar to that of nanocomposite with concentration of 5 wt% before stretching.

It is inevitable that some residues existing in neat LDPE, which may decompose under high voltage strength, lead to the heterocharges near the electrodes [33]. On the other hand, because of

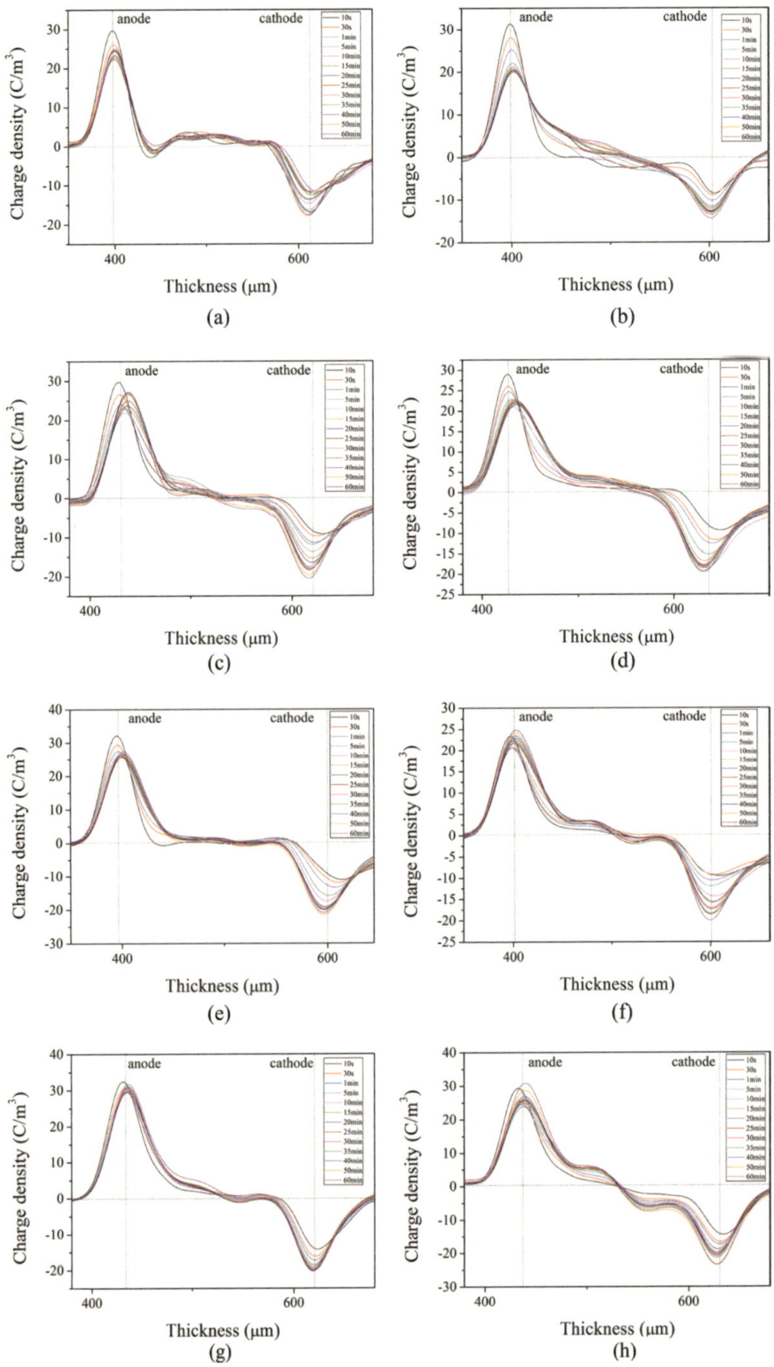

Figure 2. Space charge behavior of LDPE and LDPE/MgO nanocomposites before and after stretching during the application of a DC electric field of 40 kV/mm. (a) Neat LDPE, (b) stretching-neat LDPE, (c) 1 wt% LDPE/MgO, (d) stretching 1 wt% LDPE/MgO, (e) 3 wt% LDPE/MgO, (f) stretching 3 wt% LDPE/MgO, (g) 5 wt% LDPE/MgO, and (h) stretching 5 wt% LDPE/MgO.

the packet-like space charges observed in neat LDPE, the carriers injected by electrodes move to another electrode which may also create the heterocharges. Therefore, the heterocharges observed in this chapter may be caused by these two reasons. Due to the lack of adjacent atoms around the surface of nanoparticles, nanoparticles have high chemical activity, leading to the interface between nanoparticles and polymer matrix, which may dominate the electrical properties of nanocomposites [34]. In order to understand this interface, the multicore model has been proposed by Tanaka et al. [20]. According to this model, the interface between nanoparticles and polymer matrix can be divided into three layers: bonded layer, bond layer, and loose layer. The deep charge traps are almost distributed in bonded layer but the shallow charge traps are distributed in loose layer. Based on the research of Takada et al. [32], nanoparticle doping will increase the amount and depth of charge traps. Therefore, the carriers injected by electrodes will be trapped in charge traps leading to the homocharges in LDPE/MgO nanocomposites. On the other hand, because of the large amount of shallow traps distributed in loose layer, electrons or holes can move along the voltage field by trapping and detrapping, which means there may be overlaps between loose layers of nanoparticles as the concentration is high, such as 3 wt%, providing routes for electrons or holes and decreasing the amount of trapped carriers. This may be one of the reasons for the suppression of space charges accumulation in nanocomposite with concentration of 3 wt%. However, agglomerates may exist as the concentration is too high, such as 5 wt%, which may decrease the special surface area of nanoparticles and affect the interface between nanoparticles and polymer matrix, leading to the low capacity of suppression space charges accumulation. As shown in **Figure 2(e)** and **(g)**, there are more space charges in nanocomposite with concentration of 5 wt% in comparison with nanocomposite with concentration of 3 wt%.

3.3. Dielectric properties

Figure 3 is the dielectric properties of neat LDPE and LDPE/MgO nanocomposites before and after stretching. It is obvious that the values of permittivity of nanocomposites increase with the increase of concentration of nanoparticles, which is the same as the research of Ishimoto et al. [35]. Additionally, after stretching, the values of permittivity of all samples increase with different increasing amplitudes, in which neat LDPE has the smallest increasing amplitude but nanocomposite with concentration of 3 wt% has the largest.

Considering the large amount of free volume distributed in the interface between nanoparticles and polymer matrix, which may cause Maxwell-Wagner interface polarization [33], the values of permittivity of LDPE/MgO nanocomposites increase after adding nanoparticles. Furthermore, because the free volume mainly distributes in the loose layer of the interface, the Maxwell-Wagner interface polarization may be affected by loose layer in nanocomposites. On the other hand, the values of permittivity of nanocomposites may also increase after adding nanoparticles because of the large value of permittivity of MgO nanoparticles [35]. Additionally, according to **Figure 3(a)**, the permittivity of nanocomposite whose concentration is 5 wt% has a slope in the frequency domain from 10^{-1} to 10^{0} Hz. Meanwhile, **Figure 2(g)** shows that there is a large amount of space charges in nanocomposite with concentration of 5 wt%. Therefore, it is considered that the slope may be caused by the space charge polarization in nanocomposites.

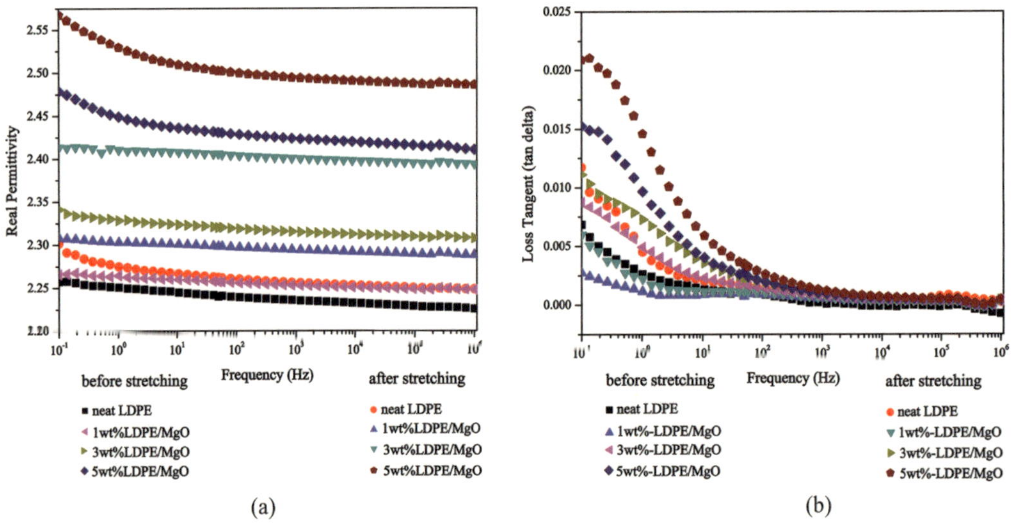

Figure 3. Dielectric properties of neat LDPE and LDPE/MgO nanocomposites before and after stretching during the application of an AC electric field of 5 V/mm. (a) Frequency dependence of real permittivity and (b) frequency dependence of loss tangent.

As shown in **Figure 3(b)**, before stretching, the gradient of slope of neat LDPE is larger than that of nanocomposite with concentration of 1 wt% but smaller than that of nanocomposites with concentration of 3 and 5 wt%. After stretching, there are little changes in the frequency domain from 10^2 to 10^6 Hz in all samples after stretching but a large increase in amplitudes in the frequency domain from 10^{-1} to 10^2 Hz, in which nanocomposite with concentration of 5 wt% has the highest value, about 0.02 at the frequency of 0.1 Hz.

Additionally, it is well known that the complicated relaxation processes of LDPE/MgO nanocomposites before and after stretching can be described by a modified Debye equation, which gives the frequency-dependent complex permittivity as follows:

$$\varepsilon^*(\omega) = \varepsilon_\infty + \left(\varepsilon_s - \varepsilon_\infty\right) / \left[1 + \left(i\omega\tau_0\right)^{1-\alpha}\right] \tag{2}$$

where α is an empirical constant with the value between 0 and 1. The case $\alpha = 0$ corresponds to the Debye model that has a single relaxation time. ε_0 is the vacuum permittivity, ε_s is the static permittivity, ε_∞ is the dielectric constant at high frequency, ω is the angular frequency $2\pi f$, and τ_0 is the mean relaxation time. Furthermore, in **Figure 3**, the decrease of ε'' at low frequency can be attributed to the suppression of electrical conduction by the nanoparticles. Therefore, the contribution of electrical conduction to ε'' has to be considered, and the term $\sigma_0/i\varepsilon_0\omega$ should be added into Eq. (3), where σ_0 represents the specific conductivity and ε_0 is the permittivity of free space. Therefore, the imaginary part ε'' can be given according to the Debye function Eq. (2) by

$$\varepsilon''(\omega) = \frac{\sigma_0}{\varepsilon_0\omega} + \left(\varepsilon_s - \varepsilon_\infty\right) \times \frac{\left(\omega\tau_0\right)^{1-\alpha}\cos\frac{\pi\alpha}{2}}{1 + 2\left(\omega\tau_0\right)^{1-\alpha}\sin\frac{\pi\alpha}{2} + \left(\omega\tau_0\right)^{2(1-\alpha)}} \tag{3}$$

When more than one relaxation peak exists in the polycrystalline ceramics, Eq. (3) can be rewritten as

$$\varepsilon''(\omega) = k_0 \omega^{\alpha_0} + \sum_{i=1}^{n} \frac{k_i \left(\omega \tau_{0i}\right)^{1-\alpha_i} \cos\frac{\pi \alpha_i}{2}}{1 + 2\left(\omega \tau_{0i}\right)^{1-\alpha_i} \sin\frac{\pi \alpha_i}{2} + \left(\omega \tau_{0i}\right)^{2(1-\alpha_i)}} \quad (4)$$

The first part of Eq. (4) is related to temperature-dependence electrical conductivity and the second part to the thermal polarization. The excellent agreement between the experimental data and the fitting curve based on Debye function was obtained, as shown in **Figure 4**.

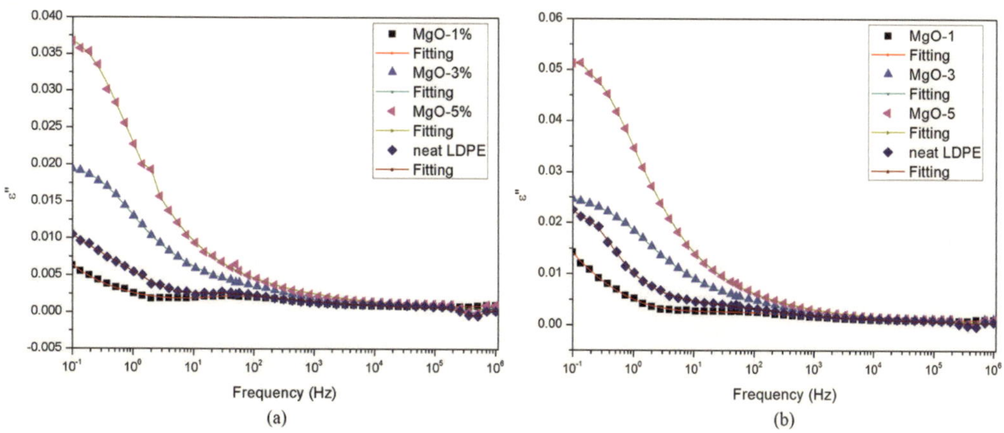

Figure 4. The representative fitting results of the dielectric loss based on Debye theory before and after stretching. (a) Before stretching and (b) after stretching.

3.4. Volume resistivity

Figure 5 is the volume resistivity of neat LDPE and LDPE/MgO nanocomposites. As shown in the image, the values of volume resistivity increase after adding nanoparticles and decrease with the increase of concentration of nanoparticles. The nanocomposite with concentration of 1 wt% has the highest value. After stretching, the values of volume resistivity of neat LDPE and nanocomposites with concentration of 3 and 5 wt% increase but decrease in nanocomposites with concentration of 1 wt%.

It is recognized that conductivity κ can be expressed by Eq. (2) [30],

$$\kappa = en\mu \quad (5)$$

where e, n, and μ represent the elementary charge, carrier density, and mobility, respectively. It is noted that the carriers are probably to be captured by traps at the interface between polymer matrix and nanoparticles in LDPE/MgO nanocomposites, decreasing the carrier mobility [35]. According to Eq. (1), the conductivity will decrease with the decrease of carrier mobility. Furthermore, when the concentration of nanofillers is high, such as 3 and 5 wt%, there are overlaps of transmission layers, increasing the carrier mobility as well as the conductivity of nanocomposites. Therefore, the volume resistivity decreases with the increase of concentration.

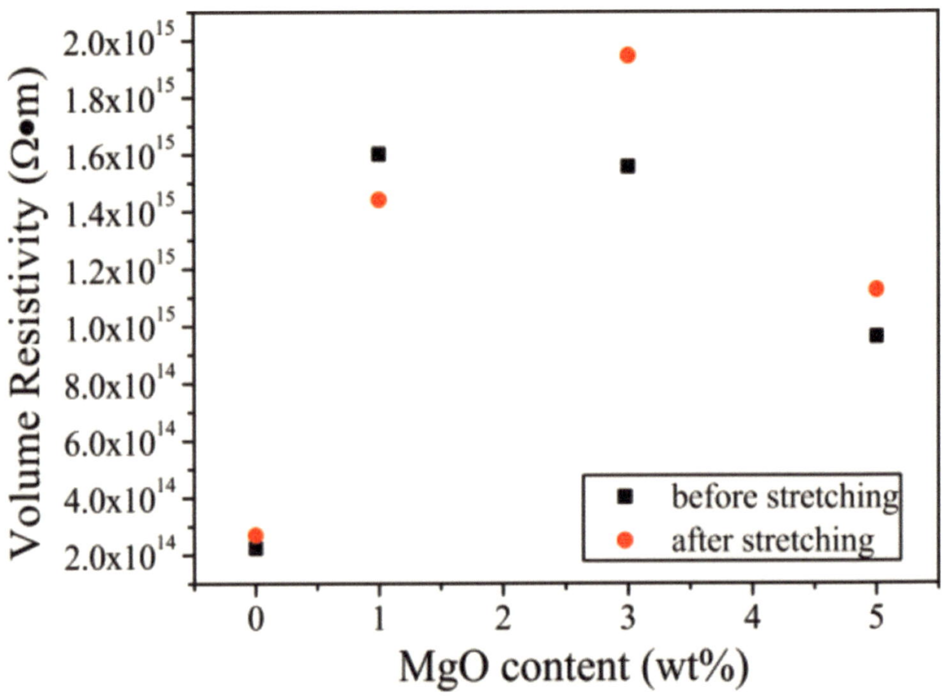

Figure 5. The dependence of MgO nanofiller content on volume resistivity during the application of a DC electric field of 2.5 kV/mm.

4. Conclusions

The effect of stretching on electrical properties of nanocomposites has been investigated. On the basis of the test results, the conclusions are as follows:

(1) The DC breakdown strengths of neat LDPE and nanocomposites decrease after stretching, which are contrary to the volume resistivity of all samples except the nanocomposites with concentration of 1 wt%. Additionally, the values of permittivity of all samples also increase after stretching as well as the loss tangent especially in the frequency domain from 10^{-1} to 10^2 Hz.

(2) The homocharges increase in all samples after stretching except the nanocomposite whose concentration is 1 wt% with a little more positive space charges accumulated in the sample. This may be caused by the increase of amorphous area as well as the change of interface between nanoparticles and polymer matrix after stretching.

(3) Because of the weakness in connection between molecules in loose layer, it is considered that the area of loose layer may increase after stretching, increasing the amount and depth of charge traps as well as the free volume. The change of loose layer may be the main reason for the change of electrical properties of nanocomposites after stretching.

Acknowledgements

This work was supported by the National Science Foundation of China (51321063).

Author details

Youyuan Wang* and Can Wang

*Address all correspondence to: y.wang@cqu.edu.cn

State Key Laboratory of Power Transmission Equipment & System Security and New Technology, Chongqing University, Chongqing, China

References

[1] G. Mazzantia, G. C. Montanari, F. Palmieri and J. Alison. Apparent trap-controlled mobility evaluation in insulating polymers through depolarization characteristics derived by space charge measurements. J. Appl. Phys. 2003;**94**: 5997–6004.

[2] C. Calebrese, L. Hui, L. S. Schadler and J. K. Nelson. A review on the importance of nanocomposite processing to enhance electrical insulation. IEEE Trans. Dilectr. Electr. Insul. 2011;**18**(4): 938–945.

[3] K. Yang, G.-J. Zhang, D.-M. Tu and Y. Zhang. Space charge and electroluminescence characteristics of thermally aged LDPE films. Appl. Surf. Sci. 2008;**255**: 2735–2739.

[4] J. Katayama, Y. Ohki, N. Fuse, M. Kozako and T. Tanaka. Effects of nanofiller materials on the dielectric properties of epoxy nanocomposites. IEEE Trans. Dilectr. Electr. Insul. 2013;**20**: 157–165.

[5] X. Huang, L. Xie, K. Yang, C. Wu, P. Jiang, S. Li, S. Wu, K. Tatsumi and T. Tanaka. Role of interface in highly filled epoxy/BaTiO3 nanocomposites. Part I-correlation between nanoparticle surface chemistry and nanocomposite dielectric property. IEEE Trans. Dilectr. Electr. Insul. 2014;**21**: 467–479.

[6] R. J. Fleming, A. Ammala, P.S. Casey and S.B. Lang. Conductivity and space charge in LDPE containing nano- and micro-sized ZnO particles. IEEE Trans. Dilectr. Electr. Insul. 2008;**15**: 118–126.

[7] P. Khodaparast and Z. Ounaies. Influence of dispersion states on the performance of polymer-based nanocomposites. Smart Mater. Struct. 2014;**23**: 1–13.

[8] R. J. Fleming, T. Pawlowski, A. Ammala, P. S. Casey and K. A. Lawrence. Electrical conductivity and space charge in LDPE containing TiO_2 nanoparticles. IEEE Trans. Dilectr. Electr. Insul. 2005;**12**: 745–753.

[9] L. Hui, L. S. Schadler and J. K. Nelson. The influence of moisture on the electrical properties of crosslinked polyethylene/silica nanocomposites. IEEE Trans. Dilectr. Electr. Insul. 2013;**20**: 641–653.

[10] K. Y. Lau, A. S. Vaughan, G. Chen, I. L. Hosier and A. F. Holt. On the dielectric response of silica-based polyethylene nanocomposites. J. Phys. D: Appl. Phys. 2013;**46**: 1–9.

[11] T. Imai, F. Sawa, T. Nakano, T. Ozaki, T. Shimizu, M. Kozako and T. Tanaka. Effects of nano- and micro-filler mixture on electrical insulation properties of epoxy based composites. IEEE Trans. Dilectr. Electr. Insul. 2006;**13**: 319–326.

[12] T. Tanaka, A. Bulinski, J. Castellon, M. Fréchette, S. Gubanski, J. Kindersberger, G. C. Montanari, M. Nagao, P. Morshuis, Y. Tanaka, S. Pélissou, A. Vaughan, Y. Ohki, C. W. Reed, S. Sutton and S. J. Han. Dielectric properties of XLPE/SiO2 nanocomposites based on CIGRE WG D1.24 cooperative test results. IEEE Trans. Dilectr. Electr. Insul. 2011;**18**: 1484–1517.

[13] S. Albayrak, C. Becker-Willinger, M. Aslan and M. Veith. Influence of nano-scaled zirconia particles on the electrical properties of polymer insulating materials. IEEE Trans. Dilectr. Electr. Insul. 2012;**19**(1): 76–82.

[14] T. J. Lewis. Charge transport in polyethylene nano dielectrics. IEEE Trans. Dilectr. Electr. Insul. 2014;**21**(1):340–351.

[15] C. Zou, J. C. Fothergill and S. W. Rowe. The effect of water absorption on the dielectric properties of epoxy nanocomposites. IEEE Trans. Dilectr. Electr. Insul. 2008;**15**(1):107–117.

[16] K. Y. Lau, A. S. Vaughan, G. Chen, I. L. Hosier, A. F. Holt and K. Y. Ching. On the space charge and DC breakdown behavior of polyethylene/silica nanocomposites. IEEE Trans. Dilectr. Electr. Insul. 2014;**21**(1): 340–351.

[17] X. Wang, Z. Lv, K. Wu, X. Chen, D. Tu and L. A. Dissado. Study of the factors that suppress space charge accumulation in LDPE nanocomposites. IEEE Trans. Dilectr. Electr. Insul. 2014;**21**(4): 1670–1679.

[18] B. X. Du, T. Han and J. G. Su. Tree characteristics in silicone rubber/SiO2 nanocomposites under low temperature. IEEE Trans. Dilectr. Electr. Insul. 2014;**21**(2): 503–510.

[19] R. C. Smith, C. Liang, M. Landry, J. K. Nelson and L. S. Schadler. The mechanisms leading to the useful electrical properties of polymer nanodielectrics. IEEE Trans. Dielectr. Electr. Insul. 2008;**15**(1): 187–196.

[20] T. Tanaka, M. Kozako, N. Fuse, and Y. Ohki. Proposal of a multi-core model for polymer nanocomposite dielectrics. IEEE Trans. Dielectr. Electr. Insul. 2005;**12**(4): 669–681.

[21] T. Tanaka. Dielectric nanocomposites with insulating properties. IEEE Trans. Dielectr. Electr. Insul. 2005;**12**(5): 914–928.

[22] N. Guarrotxena, N. Vella, J.L. Millan, S. Agnel and A. Toureille. Effect of stretching on space charge behavior of poly(vinyl chloride) as analysed by the thermal step method and the thermally stimulated depolarization currents technique. Polym. Int'l. 2000;**49**:1279–1282.

[23] G. Chen and M. R. Kamaruzzaman. Impact of mechanical deformation on space charge in XLPE. In: IEEE 7th Int'1. Conf. Solid Dielectrics (ICSD); pp. 510–513.

[24] Z.-M. Dang, S.-H. Yao and H.-P. Xu. Effect of tensile strain on morphology and dielectric property in nanotube/polymer nanocomposites. Appl. Phys. Lett. 2007;**90**: 012907.

[25] C. Feng and L. Y. Jiang. Investigation of uniaxial stretching effects on the electrical conductivity of CNT–polymer nanocomposites. J. Phys. D: Appl. Phys. 2014;**47**: 405103.

[26] S. P. Bao, G. D. Liang and S. C. Tjong. Effect of mechanical stretching on electrical conductivity and positive temperature coefficient characteristics of poly(vinylidene fluoride)/carbon nanofiber composites prepared by non-solvent precipitation. Carbon. 2011;**49**: 1758–1768.

[27] M. Urvakis, A. Kupreviciute, J. Banys, J. Macutkevic, B. Mayoral and T. McNallyb. Effect of annealing and biaxial deformation on the dielectric properties of composites of multiwall carbon nanotubes and poly(ethylene terephthalate). J. Nanophotonics. 2012;**6**: 061708.

[28] S. Li, W. Wang, S. Yu and H. Sun. Influence of hydrostatic pressure on dielectric properties of polyethylene/aluminum oxide nanocomposites. IEEE Trans. Dielectr. Electr. Insul. 2014;**21**(2): 519–528.

[29] M. Jarvid, A. Johansson, R. Kroon, J. M. Bjuggren, H. Wutzel, V. Englund, S. Gubanski, M. R. Andersson and C. Müller. A new application area for fullerenes: voltage stabilizers for power cable insulation. Adv. Mater. 2015;**27**: 897–902.

[30] Y. Murakami, M. Nemoto, S. Okuzumi, S. Masuda, M. Nagao, N. Hozumi, Y. Sekiguchi and Y. Murata. DC conduction and electrical breakdown of MgO/LDPE nanocomposite. IEEE Trans. Dilectr. Electr. Insul. 2008;**15**(1): 33–39.

[31] L. Zhang, Y. Zhou, X. Cui, Y. Sha, T. H. Le, Q. Ye and J. Tian. Effect of nanoparticle surface modification on breakdown and space charge behavior of XLPE/SiO2 nanocomposites. IEEE Trans. Dilectr. Electr. Insul. 2014;**21**(4): 1554–1564.

[32] T. Takada, Y. Hayase, Y. Tanaka and T. Okamoto. Space charge trapping in electrical potential well caused by permanent and induced dipoles for LDPE/MgO nanocomposite. IEEE Trans. Dilectr. Electr. Insul. 2008;**15**(1): 152–160.

[33] M. Roy, J. K. Nelson, R. K. MacCrone, L. S. Schadler, C. W. Reed, R. Keefe and W. Zenger. Polymer nanocomposite dielectrics – the role of the interface. IEEE Trans. Dilectr. Electr. Insul. 2005;**12**(4): 629–643.

[34] T. J. Lewis. Interfaces are the dominant feature of dielectrics at the nanometric level. IEEE Trans. Dilectr. Electr. Insul. 2004;**11**(5): 739–753.

[35] K. Ishimoto, E. Kanegae, Y. Ohki, T. Tanaka, Y. Sekiguchi, Y. Murata and C. C. Reddy. Superiority of dielectric properties of LDPE/MgO nanocomposites over microcomposites. IEEE Trans. Dilectr. Electr. Insul. 2009;**16**(6): 1735–1742.